SUSTAINABLE AGRICULTURAL ENTREPRENEURSHIP

The urban area as engine for new economic activity

The six guises of the successful agricultural entrepreneur
illustrated on the basis of nine projects

TransForum

Colofon

Authors	Anne-Claire van Altvorst Henk van Latesteijn Karin Andeweg Lia Spaans Rik Eweg Sander Mager
Copywriting	Leo Stumpel
Contributions	Blonk Environmental Consultants, Gouda, the Netherlands André Nijhof (Nyenrode Business University) and Taco van Someren (Ynnovate)
Translation	Jan Arriens
Design	Creja, Leiderdorp
Photography	See individual credits
Infographics	Schwandt Infographics, Houten
Published by	**TransForum** P.O. Box 80 2700 AB Zoetermeer The Netherlands T +31(0)79 347 09 10 F +31(0)79 347 04 04 E info@transforum.nl I www.transforum.nl/en

ISBN 978-94-90192-17-4

First edition, 2011 © TransForum, P.O. Box 80, 2700 AB Zoetermeer, the Netherlands

TransForum

Anne-Claire van Altvorst

Karin Andeweg

Rik Eweg

Henk van Latesteijn

Sander Mager

Lia Spaans

SUSTAINABLE AGRICULTURAL ENTREPRENEURSHIP

The urban area as engine for new economic activity

The six guises of the successful agricultural entrepreneur illustrated on the basis of nine projects

PREFACE

Our world is urbanizing at high speed. Since 2008 more than half of the world's inhabitants are living in cities. And this is expected to rise up to 70% or 6.4 billion people in 2050. Historically, most settlements were built in deltas. These have the most fertile soils and are known for a high biodiversity. Also these deltas are strategic locations for trade. It is for that reason that these deltas also make the best location for agricultural activities. And it are exactly these deltas where the largest and quickest urbanisation is occurring. As a result, competition for the scarce land in these regions is growing. In the mean time, more and more food is needed for the urban population.

The north-western European delta is an example of a metropolitan area that is under high pressure. In addition to the increasing urbanisation, society is making new demands for agricultural products. Agriculture needs to take into account animal welfare, the environment and the landscape. This puts agricultural entrepreneurs on a crossroads: producing more food on less available space and on a more sustainable basis to also match with consumers' expectations and preferences.

Situated in the north-western European delta is the Netherlands, one of the most densely populated countries in the world. The Netherlands is known for its highly efficient and innovative way of agricultural production. Also in the Netherlands, agricultural entrepreneurs ask themselves whether their traditional products and methods still suffice for the new demands from society. Simply producing more of the same is not a viable option. Is it possible to find new ways of production that produce sufficient and responsible food and at the same time contribute to sustainable development? Above all, the future must lie in new business models. In this respect, the Netherlands can be seen as an innovative 'pressure cooker' for developments in other parts of the world.

Over the past six years, the Dutch innovation programme TransForum experimented with new ways of agricultural production that are profitable and respect the environment and animal welfare. In the course of doing so, a wealth of experience is built up from entrepreneurs who took up the challenge to develop new more sustainable business models. The experiences from these Dutch agricultural entrepreneurs in a highly urbanised environment are a potential source of inspiration for entrepreneurs in other parts of the world wishing to take up the same challenge. As a member of the International Advisory Board of TransForum, I have seen the importance of the experiences in the Netherlands elsewhere in the world. It is of interest not just for entrepreneurs but also provides tools for 'business-minded' civil servants, politicians, research workers, students and representatives of societal organisations. The challenges faced by the agrosector are so complex as to defy solution by entrepreneurs alone. The most instructive aspect therefore concerns the way in which multidisciplinary forms of cooperation came about: with chain partners, with the public sector, with community groups and even with end-users.

I am enthusiastically recommending this book to everybody involved in taking up the challenge to innovate the agricultural sector. Anyone wanting to know more about the projects and work of TransForum should consult the organisation's website: www.transforum.nl/en.

Hans Jöhr
Head of Agriculture
Nestlé, Switzerland

CONTENTS

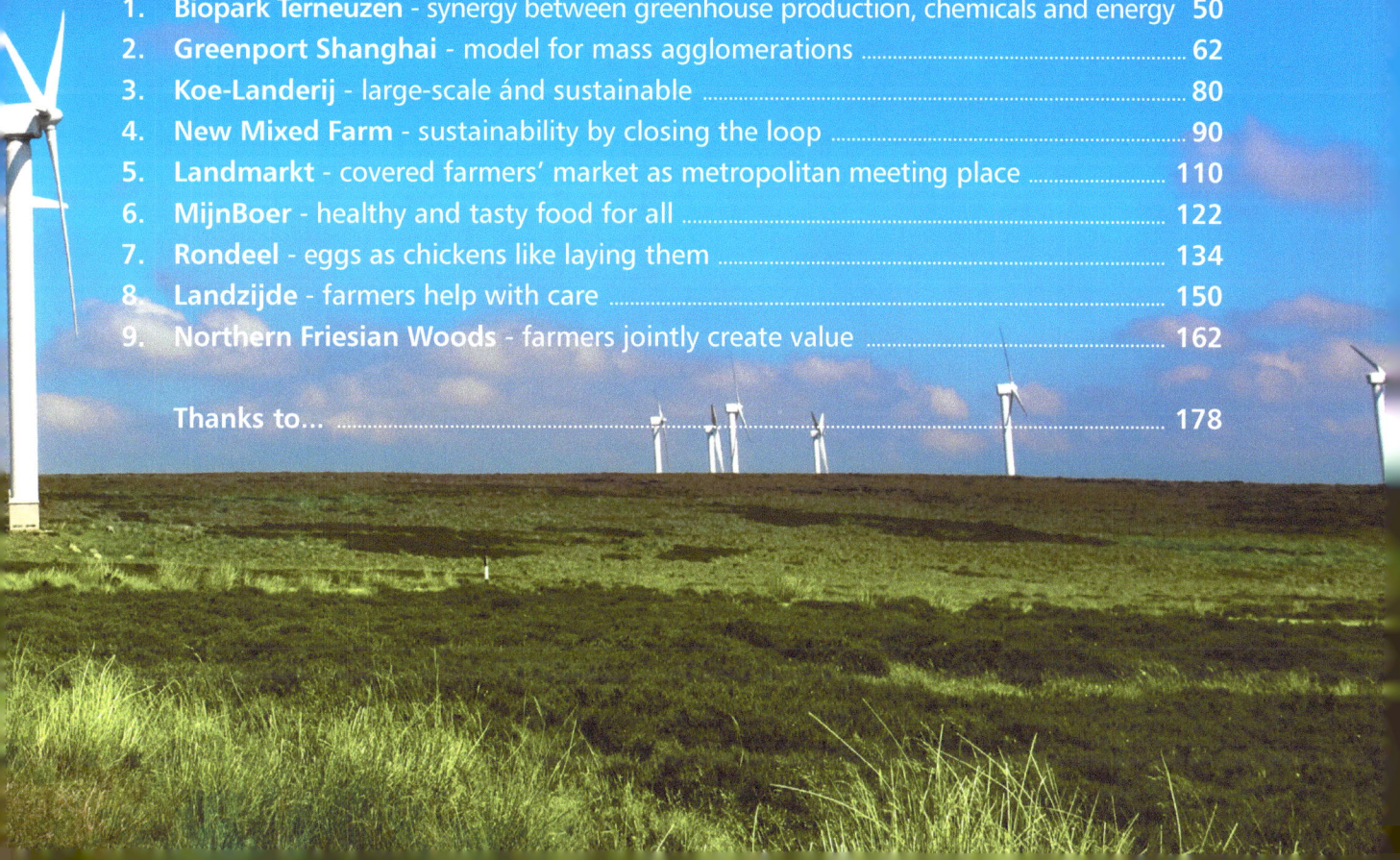

CHANGE IS IN OUR NATURE

Not so long ago we lived in a world that was dominated by agriculture at small scale. There were, of course, also cities. And there was industry and trade. But change took place at high speed ...

The northwest European Rhine delta, for example, has evolved into one of the biggest urban agglomerations in the world, providing room for living for some 30 million people. This has major consequences for the agricultural sector in this area. The farmer of today can no longer get by just by growing vegetables, cultivating fields or keeping livestock. The changing situation calls for a wider range of skills for the agricultural entrepreneur: ones outlined in this book. They are discussed and illustrated on the basis of nine innovative projects, selected from the 34 projects that TransForum has helped develop in recent years. These projects all took place in the context of the northwest European delta. Experiences of these projects provide important lessons for the sustainable development of agriculture in metropolitan areas worldwide.

TRANSFORUM'S AIMS

TransForum was set up in order to stimulate the sustainable development of Dutch agriculture. To this end, it is imperative for the agricultural system to be linked up to the urban environment. This is possible only given intensive and wholehearted cooperation among agricultural entrepreneurs, as well as with research institutes, government agencies, societal organisations and other businesses. In this way new economic activity can be developed that is profitable, respects the environment and improves the welfare of both people and animals.

Metropolitan agriculture

Virtually all the projects that TransForum has carried out in recent years support the vision that agriculture and its urbanised or metropolitan environment should reconnect with each other in the interests of more sustainable agriculture. This we refer to as 'metropolitan agriculture'.

The essence of metropolitan agriculture is that the urbanised environment in fact offers great opportunities for the more sustainable development of agriculture. Conversely, agriculture is indispensable for the more sustainable development of those urban areas.

Metropolitan agriculture covers all types of more sustainable agriculture and the related arrangement of the agroproduction chain (including agroparks, care farming and alternative forms of distribution). In all cases, the activities must take place in a metropolitan environment, are explicitly concerned with the divergent needs of the urban population and make use of the typical urban characteristics of that environment.

Ongoing interaction indispensable: adversaries become partners

The spatial and cultural division between agricultural producers and consumers has led to little mutual understanding. The way in which agricultural entrepreneurs go about their work must be consistent with the standards and values of the general public and consumers.

Ongoing interaction between agriculture and its metropolitan environment is indispensable for the more sustainable development of agriculture. This creates new links between agriculture and the city. These are a source of inspiration for innovation, are profitable, respect the environment and improve the welfare of both people and animals.

New alliances: indispensable for sustainable development

Successful agricultural entrepreneurs manage to deliver added value in new fields or activities next to traditional agricultural products and services. In doing so they contribute to a more sustainable development. This will succeed only if they continue to cooperate with allies some of whom will be new and who will come from unexpected quarters. This collaboration generates the necessary political and public base of support for their new activities. A characteristic feature of all forms of metropolitan agriculture is the new links that these activities involve. This calls for cooperation with industries and sectors that have not been traditional partners of the agricultural industry. Examples include the cooperation between agriculture (as a provider of care) and health care, agriculture (as a producer of energy) and energy companies, and fertiliser manufacturers (as suppliers of CO_2 and residual heat) and horticulture.

Cooperation with governments and societal organisations is also required: innovations nearly always run into obstacles as they do not fit in with the existing legislation and regulations. In this regard active participation by governments can provide the necessary way-out. For an entrepreneur to recoup his investments in innovations aimed at greater sustainability, there must be customers who can spot the added value and who are prepared to pay for it. Support by societal organisations can provide those customers with just the nudge they need. This calls for close cooperation by parties who may never have dealt with one another before and who may even have been used to seeing each other as opponents. Successful innovation in the field of sustainable agriculture nearly always involves close cooperation among knowledge institutions, governments, societal organisations and the private sector.

If it is our ambition to provide encouragement on a larger scale for agricultural innovations that will lead in practice to a more sustainable development of the agrofood sector, an agro-innovation system will in our view need to be set up collectively. A coherent system that eliminates the barriers on which innovations currently often founder. In such a system a role needs to be provided not just for research institutes, government agencies, societal organisations and businesses but quite clearly also for project developers and investors.

HOW AND WHAT

This book is based on the learning experiences gained from 34 TransForum innovative projects conducted between 2004 and 2010. These projects brought together entrepreneurs and researchers, government agencies and representatives of societal organisations working on innovations in the agrosector and in green space. Nine of these projects are selected for more detailed description from the viewpoint of entrepreneurs.

This book is therefore a source of inspiration for entrepreneurs who wish to invest in sustainable development on a profitable basis.

A book for entrepreneurs

The most important audience for this book are entrepreneurs. This does not mean that the book will be of no interest to others: the development of sustainable new activities in the agrosector is also a challenge for researchers, administrators, civil servants and representatives of societal organisations. Finally, the book will also be of interest for students: the innovators of the future.

How to read this book?

Part I describes the challenges faced by agricultural entrepreneurs and the wide range of skills they need in order to invest profitably in sustainable development.

Part II goes on to examine the nine innovative projects. First of all a brief account is provided of what makes each of these projects such an interesting example of new and sustainable entrepreneurship. This is followed by more detailed descriptions.

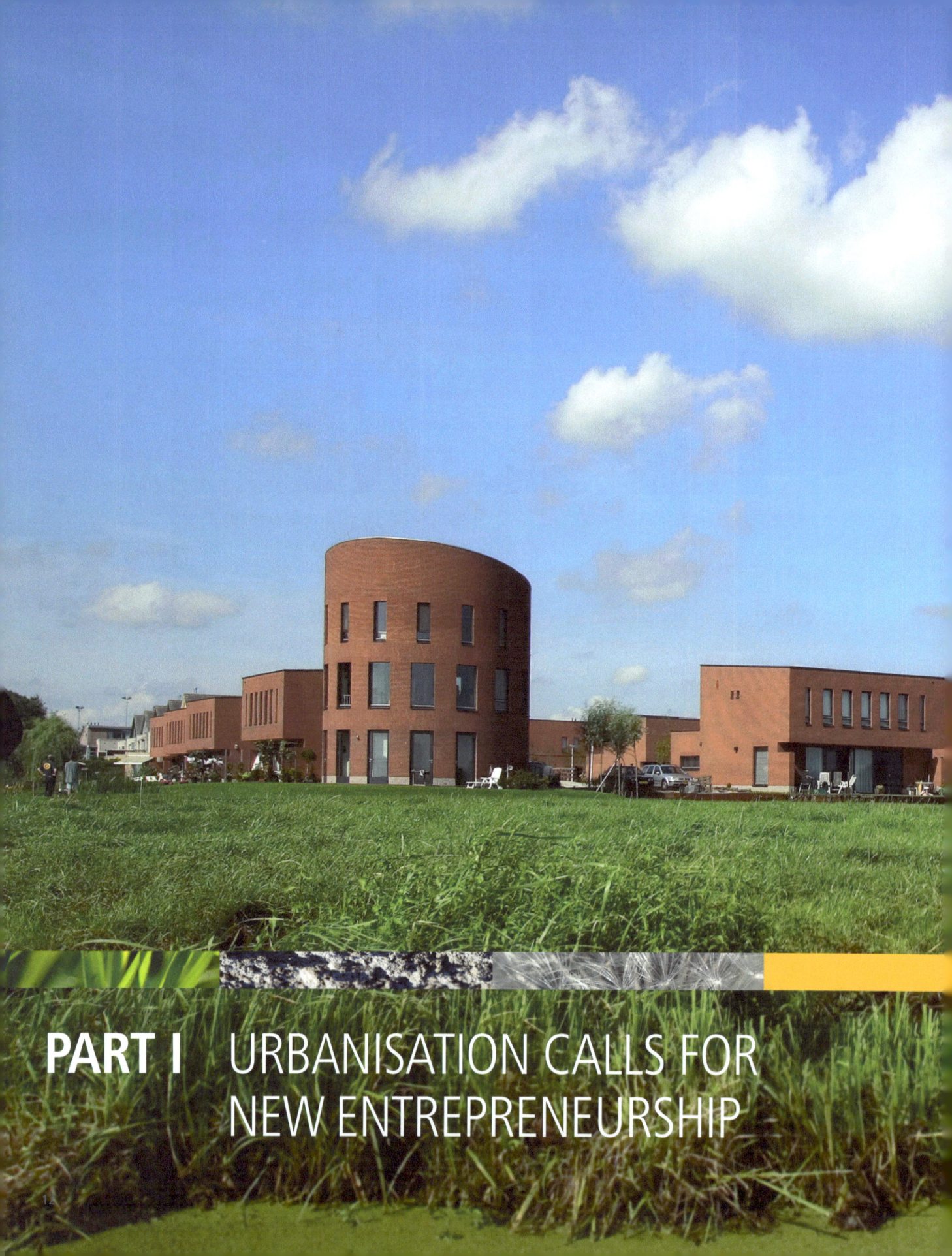

PART I URBANISATION CALLS FOR NEW ENTREPRENEURSHIP

1. THE CHALLENGE FOR THE NEW AGRICULTURAL ENTREPRENEUR

A CHANGING ENVIRONMENT CALLS FOR A DIFFERENT METHOD OF FOOD PRODUCTION: METROPOLITAN AGRICULTURE

In recent decades the deltas in our world have evolved from agricultural areas with towns and cities into urbanised regions with green space. This we call a metropolitan area. Such areas have far-reaching consequences for the type of agriculture that can be conducted; new forms of activity are required that are more closely attuned to the demand in an urban environment. We call this new way of agricultural production Metropolitan Agriculture.

The Dutch agricultural sector remains one of the most important exporters of agricultural products in the world. In contrast to many products made in other sectors of the economy, these are all products that are produced in the Netherlands and processed into end-products. The strength of the Dutch agricultural sector is due in particular to its strong national and international networks and its exceptional in-depth knowledge. This makes the Dutch system of agricultural production an outstanding international example of knowledge-intensive production in a densely populated delta region.

The problem

The increase in urbanisation has also had major consequences. City and countryside have grown further apart in recent decades. In the drive to increase production and make efficiency gains, agriculture has over time become ever more specialised and increasingly separated physically from the city, while on the other hand the modern urban dweller seldom has a realistic picture of what modern agriculture looks like and how their food is produced. There has consequently been an increase in public opposition towards increases in scale and intensity, and debates concerning the spread of animal diseases, odour nuisance, environmental pollution, animal welfare and the degradation of the landscape are the order of the day. On top of that there is the competition for agricultural land for housing, employment and recreational purposes. At the same time farmers are grappling with profitability problems, and many have difficulty keeping their heads above water. Not without reason more and more farmers are leaving the industry.

The challenge

In order to improve agriculture's image, agriculture will need to respond more effectively to the changing nature of demand, and this too provides calls for more sustainable agriculture.

TransForum sees major opportunities for agricultural entrepreneurs, particularly if we recognise that town and countryside have become even more indissolubly bound up with one another than before. They depend critically on one another: as supplier, customer, co-user of space and producer and processor of waste substances.

By establishing new links the entrepreneur is able to cut costs, work more efficiently and obtain added value from his or her products. Investments in sustainable development are not however easily recouped; this calls for a new type of entrepreneur.

Action needed in order to create new opportunities

'Responding in order to survive' will no longer suffice for the new agricultural entrepreneur. The entrepreneur of the 21st century needs different skills in order to to make successful investments in sustainable modes of production. In order to capitalise on opportunities, new business models must be developed. These are models that combine the sustainability themes of people, planet and profit (3P). The development of 3P business models requires not just new strategies but, in particular, new forms of cooperation as well.

The choice of partners with which to collaborate will depend on the entrepreneur's goals and his or her operating environment. This may concern collaboration with fellow farmers, with other industries operating in the same urban environment, with governments and with societal organisations and research institutes. This may involve working with partners with whom they previously had little if any contact or who were even adversaries. While doing so involves making an effort, cooperation has consistently shown its merit in all sorts of projects.

Three strategies for 3P business models

We have identified three strategies for the realisation of 3P business models in the agricultural sector:

- **SUSTAINABLE INTENSIFICATION: Cooperation to establish a base of public support for new intensive methods of production**
  Despite the opposition among members of the public towards agriculture, the urban population will still need

to be fed in the future. For reasons of food security, reliability, public health and certainly also animal welfare, food will preferably be produced close to cities. The growing world population and rising global living standards mean that even more food will inevitably be required in the future. This calls for both an efficient and a more sustainable method of production in order to generate public acceptance and appreciation.

The innovative projects based around this strategy are Biopark Terneuzen, Greenport Shanghai, Koe-Landerij and New Mixed Farm.

• SUSTAINABLE VALORISATION: Cooperation with new chain partners to open up existing markets

The current trends in the agrosector are the dominance of retailers, the increases in scale in agriculture, the narrowing of the fresh produce range (with fewer specialties and the focus on commodities) and price as the leading mechanism for remunerating producers. In this system of fierce price competition, quality and sustainability barely manage to get a look-in.

Under this strategy, sustainability is viewed as a new positioning opportunity. Innovative links need to be set up among chain partners, whereby sustainability is a win-win situation instead of an additional cost factor. This calls for the development of new, sustainable chains.

The innovative projects based around this strategy are Landmarkt, MijnBoer and the Rondeel.

• SUSTAINABLE DIVERSIFICATION: Cooperation for new products and markets

Successful agricultural entrepreneurs manage to deliver added value in new areas of activity other than traditional agricultural products and services. Care farming, for example, concentrates on the need for peace and quiet, spaciousness and greenery. The innovative entrepreneur is able to capture a position in new markets by bringing about the cross-fertilisation of various sectors and the development of relevant products and services. This may involve cooperation with, in some cases, totally new partners.

The innovative projects based around this strategy are Landzijde and the Northern Friesian Woods.

2. THE SIX GUISES OF THE SUCCESSFUL AGRICULTURAL ENTREPRENEUR

NOBODY IS PERFECT, BUT A TEAM CAN BE THERE!

Nine projects are described in Part II. We have much to learn from these projects. In these nine innovative projects (and indeed also in the 25 other projects) we see that the entrepreneurs need to assume a number of different guises for the successful development of their 3P business models. These guises turn out to be much the same in many projects.

The consistently required guises are described below. A new, successful approach towards agricultural entrepreneurship requires the development of sustainable business cases in a metropolitan setting. Or in other words, the necessary flexibility to adapt successfully to future requirements.

Needless to say, no one is capable of carrying off all these roles with equal success. The smart entrepreneur, however, knows how to surround himself by others with the strengths he lacks.

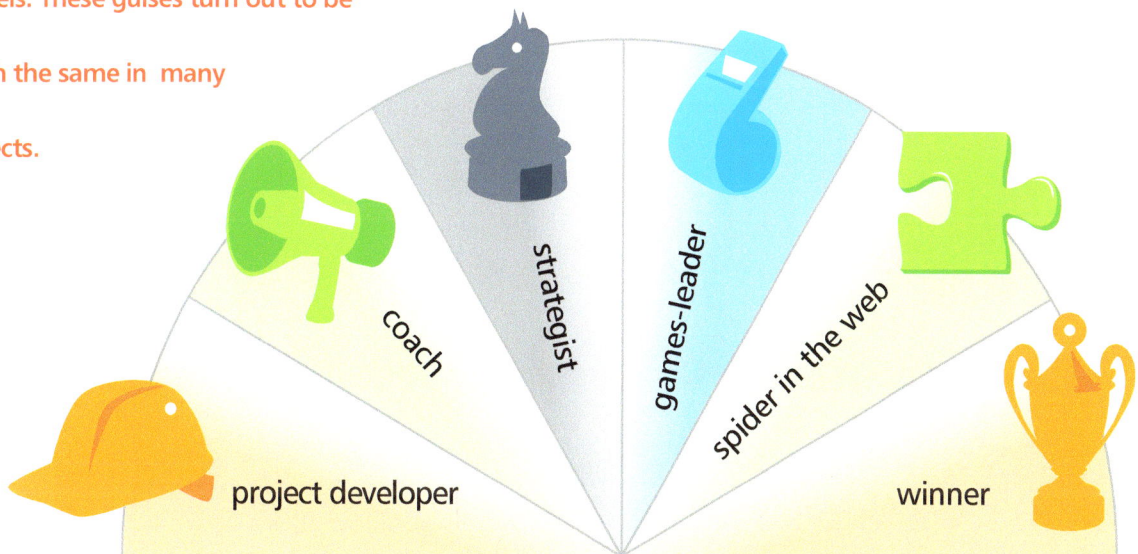

coach

strategist

games-leader

spider in the web

project developer

winner

1. The entrepreneur as project developer

An essential precondition for any new business is that it must be profitable. Investments in sustainable development are also concerned with improving the environment (planet) and social quality (people). An innovation will be successful if the new method of working gains acceptance in the market. Each innovation involves three phases, which will often overlap in practice: the planning phase, the investment phase, and the operating phase, when the return on investment in people, planet and profit (3P) is achieved.

- The result of the *planning phase* is a 3P business plan that carries conviction for investors, results in public acceptance and persuades governments to support the innovation, for example by granting licences and permits.
- In the *investment phase* the plan is converted into an actual business when entrepreneurs and public and/or private investors invest money and energy in the realisation of the plan.
- In the *operating phase* the business becomes operational, the investor obtains his projected return and value is created on people and planet aspects.

Lessons from the projects

- The planning phase often generates a clear objective and is aimed in particular at creating the confidence that the end-result will be jointly achieved. A reality check and sustainability scan are important at this stage. During the investment phase the ideas are worked out in detail and genuine commitment is sought from the parties in the form of shared investment.
- The parties involved may have different roles in the different phases: a party providing input in the planning phase may turn out to be a business partner, regulator or licensing authority in the investment phase.
- In order to speed up the investment phase it is a smart move to involve potential investors in the planning phase as well.
- In the planning phase innovation is often supported by subsidies. This entails the risk that plans will be developed that remain critically dependent on subsidies in the investment phase and (therefore) fail to provide a sufficient return on capital.
- The investment phase is marked by the fact that 'he who pays the piper calls the tune'. The most effective course of action for parties cooperating in the planning phase and wishing to remain involved in the investment phase is to help share the investment costs.
- Investments in 3P business cases may come from both public and private sources.

2. The entrepreneur as coach

Genuine breakthroughs come about when people from different backgrounds get together and develop new knowledge: entrepreneurs, civil servants, researchers and representatives of societal organisations. These parties have established roles and relationships in everyday life in relation to each other: they may for example be each other's financier, competitor, regulator, licensing authority or even opponent. They also each have their own 'remuneration mechanisms': money, power and status, the reactions of donors and members, scientific publications. These differing

roles and types of reward hamper the creative thinking process.

For genuine innovations, space to experiment needs to be organised. This may take place as follows:

- By (as coach) forming a team consisting of individuals drawn from various parties who need not represent those players 'formally' but who will have the common goal of winning the contest.
- By setting up projects or programmes in which existing political, economic and social reward structure trade-off mechanisms have been (temporarily) suspended.
- By agreeing clear rules of the game with one another (see games-leader).
- By means of creative workshops and by learning from other inspiring initiatives.
- By obtaining a time- and place-limited dispensation from the existing rules from the government.

Lessons from the projects

- When licences are awarded, the technology is assessed in terms of best proven practices. New technologies do not always meet this standard. It is desirable for the regulations to afford freedom to experiment, combined with (scientific) monitoring.
- Long-term experimentation enables all the parties to learn from a new development, which becomes an 'experimental garden'. The new knowledge allows them to realise their 'reward' within their traditional environment: money, power, status, influence (networks) or publications.
- Reflection helps them to stand back. In order to respond effectively to social developments it is necessary to take a broad view at all times and to step beyond the confines of one's own environment. New insights are generated by sparring with one another and with other innovative projects.

3. The entrepreneur as strategist

Metropolitan agriculture is not based around a single strategy. In order to capitalise on the new opportunities offered by metropolitan agriculture, at least three strategies are possible:

1 **Sustainable Intensification:** food production in metropolitan areas will need to develop new and efficient methods of production. These will be concerned with higher output per hectare or production on a bigger scale on a more sustainable basis. The new methods of production must gain public acceptance and approval. The aim is literally and figuratively to convert NIMBY (Not In My BackYard) into YBMIN (Yes, Be My Innovative Neighbour).

2 **Sustainable Valorisation:** agricultural entrepreneurs establish new chains linking up urban markets in the city with agricultural production. In the new chains they themselves share responsibility for marketing their products, and the gap between producer and consumer is narrowed. Quality and sustainability therefore obtain a more important place in the chain. The drivers of integration in the chain are assigning a value to quality and sustainability, direct contact between city and countryside, shortening the chain and a 'fairer' distribution of value within the chain.

3 **Sustainable Diversification:** by responding to the new needs of city dwellers, agricultural entrepreneurs are able to develop new products and markets. One example is care farming, in which agricultural entrepreneurs respond to clients' need for peace and quiet, spaciousness and greenery.

Lessons from the projects

- New products can also be publicly funded if they focus on public values, such as 'a valuable cultural-historical landscape'.
- A change in the chain requires various parties to change. A stepwise approach works best here.
- A choice in favour of continuing to produce for the commodity market imposes a limitation on the available resources that are required for any additional investments in sustainability. Investments in sustainability can then be financed only if they also result in lower production costs.
- The added value of a more sustainable product can be recouped if the benefit is visible to the consumer, either by means of a new formula or by being awarded a sustainability hallmark. It is important to create a base of support among societal organisations and governments at an early stage of concept development. Sometimes it can also be helpful to obtain scientific evidence of the added value by means of substantiated statements by research institutes.
- Sustainability is a very broad concept. Creating added value on the basis of sustainability alone is not straightforward. Sustainability needs to be combined with other types of added value (experience, education, recreation, transparency, taste). It is important to set up a business model that cannot be copied.
- Each new or existing chain also needs a chain orchestrator: the player who accepts responsibility for the organisation and integration of the chain.
- A professional approach is essential for all three of these strategies. That may sound self-evident, but is not always the case. Professionalisation means the development of professional expertise (knowledge, skills and attitudes) and entrepreneurship (the capacity to set up an organisation to take advantage of new market opportunities).
- New strategies may require new competencies that have to be outsourced, such as design, marketing and communication.

4.   The entrepreneur as games-leader

An innovation process is unpredictable, and so its substance cannot be agreed in advance. Agreements can, however, be reached concerning the process itself. These agreements must be accepted by the participating parties. In order to realise a new business model with various stakeholders it is important for them to trust one another, to be sure about each other's responsibilities and risks and to be clear about the goal that is to be jointly achieved.

An entrepreneur must therefore ensure that the following[1] are in harmony with one another:

- **Trusting:** the participants trust each other, don't begrudge each other their successes, reach agreement that the process provides a 'safe' environment for all participants, seek each other's reactions, have the flexibility to adjust their own position or to accept other points of view.
- **Sharing:** each participant has a story describing his own input and goals and is prepared to share that story with the other partners.

[1] *Inspired by the book Blue Ocean Strategy (W. Chan Kim and René Mauborgne, 2005)*

- **Expecting:** Proper agreements are reached concerning the expectations for the outcomes of the process. All the participants are involved in the decisions, and each one is invited to issue a response to them.

Lessons from the projects

- In a 3P business case ownership will often involve multiple and mutually dependent players in the chain. This calls for trust among these players, especially in the investment and operating phase. By enquiring specifically in the planning phase about each other's values and expectations it becomes possible to build up trust at an early stage. Start the process by formulating a shared ambition and each other's expectations.

- An open mind is required. Instead of positions adopted in advance, a broad outlook is needed: talk with external experts, organise excursions, learn from other initiatives and pay heed to developments in society.
- Commitment is crucial: the parties will only display perseverance in developing a business case if they have a direct interest in it.
- The development of an innovation requires the process to have flexibility: the freedom and willingness to change strategy and respond to new issues and opportunities. Don't think immediately in terms of solutions.
- As entrepreneur maintain the initiative in the project and reach clear agreements on each party's role, time commitment and contribution.
- Organise reflection on the method of cooperation, with someone from outside holding up a mirror to the participants.

5. The entrepreneur as spider in the web

For a 3P business model to succeed, meaningful links have to be organised. Meaningful links mean that the parties will feel genuinely involved in the innovation as they have a clear interest in it, for example because information, knowledge or money is being exchanged.

Links exist between various parties:

- Between the entrepreneur and the government: the government must provide room in its policies and licensing procedures and must participate: *'How can the government help me?'*
- Between the entrepreneur and research institutes: cooperation is instructive and leads to new insights: *'What can research institutes teach me?'*
- Between entrepreneur and society: without legitimation by society it is not possible for a business model to be realised: *'What (social) value am I creating?'*

- Among entrepreneurs themselves: cooperation with other entrepreneurs can lead to stronger business for all the players: *'With whom can I generate added value?'*

Lessons from the projects

- Involving non-governmental organisations is often difficult, but if it succeeds they can play an important role in the acceptance of the method of production, location and the product itself. In doing so they help achieve a higher price for the product.
- In order to develop a successful business model for an entire area, it is essential for the entrepreneurs to organise themselves.
- Depending on the mutual relationships, consideration may be given to placing the task of realising meaningful links with a neutral person or party.

- A clear and communal 'story' is an important binding factor in the development of a working network. It can also help to have good images, ambassadors or a striking scale-model of the intended result.
- Analysis of flows and cycles can lead to new partners for whom the closing of loops creates a shared added value.
- Learn from others' successes and failures.

- A strategic plan for clear communication to the general public is important in order to generate public support. Take the – often emotional – objections seriously. Emotions are not generally overcome by means of factual, rational information.
- The establishment and management of external networks is important in the development of innovations as they can be helpful at the operating phase.
- Seek contact with research institutes and look specifically for researchers who could take the innovation forward by means of applied research.

6. The entrepreneur as winner

Sustainable development means that people, planet and profit aspects are mutually reinforcing: this is the essence of the new business model. The essential precondition is that the investments with regard to planet and people can be recouped by saving costs and/or creating added value.

Questions that can arise in the case of a 3P business plan are:
- *'Where can planet and profit and people and profit reinforce one another?'*
- *'What value am I creating as entrepreneur with regard to people, planet and profit and for whom am I doing this?'*
- *'How can I earn money from this as entrepreneur?'*

In the planning phase the entrepreneur draws up a 3P business plan, including a communication plan and a financing plan.

Lessons from the projects
- Reflect on the risks. Can the cost of additional finance be recouped and, if not, how can this be covered (e.g. by a guarantee)? What is the non-performance risk and is a back-up system required? For the investment to be profitable a certain volume will often be needed.
- Determine whether the innovation is concerned with the development of specialties (i.e. products with an added value and commanding a higher price) or commodities (i.e. products for the world market at appropriate prices).
- Consider what dependence on subsidies might mean: government policy and grants can be unpredictable.
- Underpin the sustainability story with hallmarks, benchmarks or sound arguments, or the story will rapidly be seen through.

A NEW APPROACH TOWARDS AGRICULTURAL ENTREPRENEURSHIP IS REQUIRED

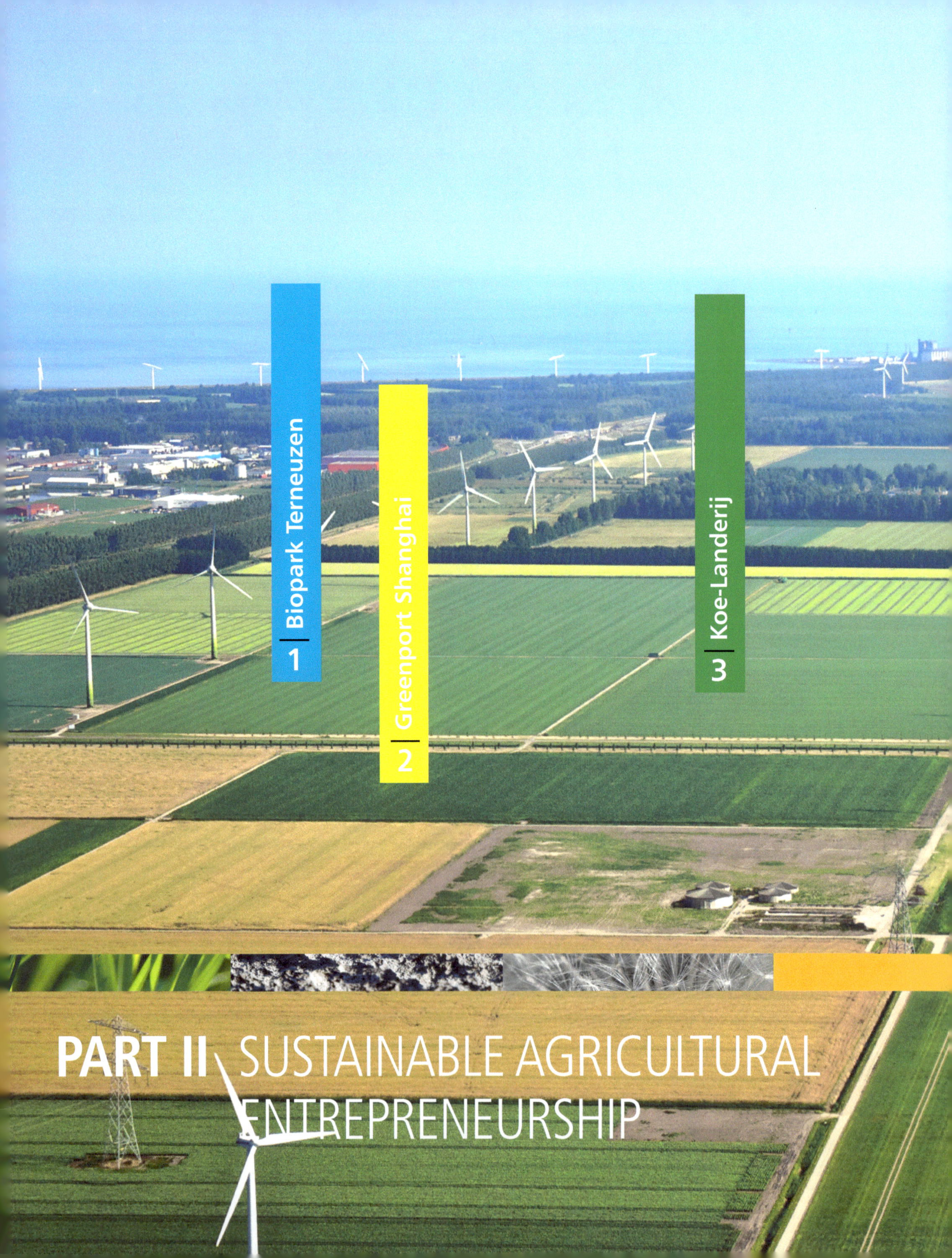

1 | Biopark Terneuzen

2 | Greenport Shanghai

3 | Koe-Landerij

PART II SUSTAINABLE AGRICULTURAL ENTREPRENEURSHIP

4 | New Mixed Farm

5 | Landmarkt

6 | MijnBoer

7 | Rondeel

8 | Landzijde

9 | Northern Friesian Woods

NINE VOYAGES OF DISCOVERY TO THE FUTURE

In the following chapters you will meet agricultural entrepreneurs who have demonstrated that a more sustainable agricultural entrepreneurship in the 21st century can be successful. They have developed new insights, from which you can benefit. TransForum has drawn up a summary of the cooperation and value creation models, of the skills you need as an entrepreneur, and of the various guises you must be able to adopt in order to be successful. But as in the case of any skill, it is the user who determines whether and how the final result is successful!

What all these entrepreneurs have in common is that they all regard the new demands made on them by a rapidly changing society not as problems but as opportunities. Whether this concerns food quality, sustainability, animal welfare or landscape preservation and management, there will always be ways of using these to improve the operational performance.

The path that has to be followed is a voyage of discovery to the future of the agricultural enterprise. It is a path that is littered with obstacles, but also stirring experiences – and often throws up surprisingly varied travelling companions.

We will meet a supplier of housing systems for laying hens who comes up with such an animal-friendly system that an animal welfare organisation is now advertising it. And a farmer/teacher who comes up with a way of restoring self-respect for urban care-patients while at the same time preserving small-scale farming and the associated landscape in the vicinity of Amsterdam. We see how challenging it is to scale up an old-fashioned, mixed farm into a modern large and sustainable system. And how a way is found for bringing the products of individual farmers directly to the customer as branded goods.

Regional pride in the rural Province of Friesland results in the preservation of the landscape and value creation, while in the Province of Zeeland synergy is created between glass horticulture, the chemical industry and energy generation. In Israel, two brothers hit upon the idea as to how large-scale dairy farming could be acceptably organised. And we meet a person who transforms the romantic picture of the large, covered food markets we associate with southern European towns into covered farmers' markets on the edge of the city of Amsterdam, where good food is offered in combination with education and entertainment.

Finally there is the most distant voyage of discovery in this book: that to Shanghai, where Dutch agricultural expertise and know-how ensures that the local government's ambitious plans can result in a fully closed system of high quality food production, sustainability, a positive energy balance and profits for the entrepreneurs.

Terneuzen Docklands

1. BIOPARK TERNEUZEN

The Ghent-Terneuzen region, on the boarder of the Netherlands and Belgium, is an important delta agglomeration: one of the most highly developed in the world, with many people, a lot of agriculture and a chemical industry. The idea of creating something totally new there is another example of the voyages of discovery in which TransForum has been involved in recent years.

The keyword in this regard is *industrial ecology*, or in other words the sustainable development of the entire area. Business consultant *Mark van Waes* of the consultancy firm *Van de Bunt* came up with the idea; *Zeeland Seaports* took it up. Mark van Waes played an important role by establishing links between the industrial companies in the region and greenhouse horticulture, based on his vision of the potential for industrial ecology, the linkage of individual producers' streams and his roots in this region. Zeeland Seaports then took over the baton.

Zeeland Seaports – the port authority of the Province of Zeeland – saw an opportunity to work together with a number of municipalities in order to link up their task of job creation with the province's policy objective of developing agroproduction parks. An important factor in this regard has been that Zeeland Seaports lays down certain stipulations for greenhouse growers wishing to set up in Biopark Terneuzen, namely that they are not allowed their own combined heat and power plants. This in turn guarantees that Zeeland Seaports will have an offtake with the new company WarmCO$_2$ that facilitates the selling of industrial heat and CO$_2$ to the greenhouse growers.

The initial 125 hectares of greenhouses are now under cultivation. The greenhouse growers are at present procuring heat and CO$_2$ from the fertiliser manufacturer Yara. The necessary pipelines have been laid and WarmCO$_2$ is operational. The biodiesel plant has been completed, while a biomass plant is under construction. Biopark Terneuzen is already a fully operational agropark on a substantial scale. What has been achieved in a period of just three years is something to be proud of.

GREENPORT SHANGHAI

Artist's impression of Greenport Shanghai on Chongming Island (Masterplan Greenport Shanghai)

2. GREENPORT SHANGHAI

Shanghai is on the mouth of the *Yangtze* River. It has an annual discharge of 59,920,000,000 m³ of freshwater. The availability of water and the fertile subsoil render the Yangtze estuary highly suitable for agriculture. But this is where Shanghai, one of the biggest metropolises in the world, is located. In terms of population it is the largest municipality in the world; with 16.7 million people (in 2000) it has an average population density of 2,657 per km². Just consider the food needs of all these Shanghai people …

Urbanisation has squeezed agriculture aside. When one thinks of Shanghai one thinks of imposing high-rise buildings, busy roads and just the odd patch of greenery. Chongming Island is the last green and largely agricultural district of Shanghai. The lack of rapid transport links to the city of Shanghai has shielded the island from urbanisation. Even now that a tunnel-bridge has been constructed,

Dongtan remains agricultural, as it has been zoned by the Chinese government for *eco-city development*.

Dongtan wishes to keep urbanisation at bay and to concentrate on green and agricultural functions for the city of Shanghai. The area is designed to become the green, recreational lung of Shanghai, as well as a warehouse for various agricultural products. This will call for a huge innovation drive in agriculture, which is of low quality, supply-driven and poorly organised. The food scandals of recent years provide even further evidence of the need for such an innovation drive.

Greenport Shanghai is a concrete plan for conducting such a drive. By *integrating* intensive agricultural production with processing and logistics and *combining* this with high sustainability performance, Greenport Shanghai will be capable of offering high volumes and unique quality, so that the rest of the island can be used primarily for landscape-based recreational purposes. The linkage of waste flows, energy generation and water purification also mean that Greenport Shanghai will fit in excellently with Dongtan's ambitious goals. On top of that, the innovativeness and transparent management of Greenport Shanghai will also make it an educational and recreational draw card.

Greenport Shanghai is still just a plan. The big challenge will be whether entrepreneurs are capable of putting this innovation into practice and what new roles they will need to play in order to do so. Or will such an ambitious plan never get off the drawing board?

Artist impression open-plan cowsheds
(Architectuurstudio SKETS, Groningen)

3. KOE-LANDERIJ

Keeping the Rhineland – the metropolitan area in the triangle of Amsterdam, Brussels and Cologne – supplied with dairy products requires high volumes of production. In doing so the *Wilms brothers* are picking up the classical role of agriculture: making sure of the supply of food for the cities in their vicinity. At the same time they seek to meet aspirations concerning animal welfare, the environment and the landscape: values that have risen to the top of the agenda with the growth in urbanisation in this region.

The Wilms brothers want to set up a business that is 'sound'. Sound in the sense that it forms part of the community and is valued by and of significance to the community. Sound because it pays attention to animal welfare and the landscape, because it makes it possible for entrepreneurs to earn a living and because it allows employees to work normal hours and to develop themselves.

The Wilms family farm goes back to 1880 and the family therefore forms part of the local community (being represented on various committees, the municipal council and in voluntary associations, etc). They regard the fact that their farm is also a valued part of the community as self-evident. They did however actively seek out inspiration and possibilities for developing their knowledge.

A visit to Israel in 2008 aroused their interest in keeping cattle in loose housing on natural soil. Their dream was to set up a large-scale dairy farm, in which the cows would be held in herds of 60 animals, divided over eight open-plan cowsheds on compost bedding. This would be an increase in scale that took into account both animal-welfare and the landscape, and that was embedded in and valued by the community.

The Wilms brothers approach matters with great curiosity and are continually looking for and open to new ideas, with a clear goal in mind, working effectively in teams with carefully selected advisers. They have included researchers and consultants in the field of communication. They also involved the municipality and province in the project team to help with the planning right from the outset.

Substantial investments are being made in communication, with the neighbours and local villagers and entrepreneurs and now also with national societal organisations. Researchers are finding it pleasant and interesting to work with the farmers and so put innovative concepts into practice. The municipality and province regard the initiative as a possibility for giving the agricultural economy in their region a fresh boost. Where possible entrepreneurs incorporate the feedback they receive into their design.

The business plan is now complete. The entrepreneurs are on the point of identifying investors for their initiative, while the municipality is ready to help them with the necessary procedures.

Artist's impression: bird's-eye view of New Mixed Farm
Bio EnergyPower Plant and poultry farm (left) and pig farm
(T R Z I N bv, illustration by Erik Visser)

4. NEW MIXED FARM

The mixed farm is of course the most classical form of farming. Arable farming, horticulture, beef cattle and dairy cattle were all combined. Waste was composted, and if necessary the entire family could enjoy the warmth from the cowshed.

The fact that you can in principle scale up this highly sustainable form of agriculture and dress it up in a modern jacket is in fact a logical translation of something that is essentially very sound but should be much better again in the modern age. This would no longer mean a single mixed farm but large, specialised businesses making use of each other's waste and residual streams – all this made possible by the large scale of the collaborating businesses, and creating the potential for substantial sustainability gains.

This vision in turn fired the enthusiasm of three leading entrepreneurs, *Peter Christiaens, Martin Houben* and *Marcel Kuijpers*, who are aware that they have to move in line with society's shifting requirements if they are to stay at the top. Together with the Province of Limburg, the Municipality of Horst, *KnowHouse* and TransForum, they set to work full of energy. Later *Huub Vorsten* and *Gert-Jan Vullings* came in, with a view to building a magnificent agropark of the future.

Unfortunately a lot of time was lost and ambitions had to be scaled down during the developmental years, particularly when an anti-industrial agricultural movement came to the stage and much energy was lost in polarizing debates.

Nevertheless, a great deal of progress has been made and the project is now almost at the implementation phase. This will make it clear in the near future that these entrepreneurs are not just closing loops and promoting animal welfare but are also able to compete keenly in the world market.

Artist's impression of Landmarkt

5. LANDMARKT

Urban dwellers have a growing demand for 'authentic', tasty food that is produced as it should be. From their holidays in Southern Europe, they are familiar with village markets where local farmers, cheese makers, butchers and bakers offer their wares. Food that tastes and smells good, looks attractive and is sold in circumstances where you can meet the people who made it.

The idea came from *Jan Boone*, who accumulated starting capital from activities in the waste-processing industry, and *Harm Jan van Dijk*, who had been in marketing with the large food processing companies Mars and Sara Lee/Douwe Egberts. They took their idea of a modern marketplace where the consumer can meet the producer on a voyage of discovery.

And within a short period of time the first Landmarkt opened on the outskirts of Amsterdam. Here, the city people and their children are able to buy genuine food from the people who make it themselves.

MijnBoer connects producer and consumer

6. MIJNBOER

Consumers of fruit and vegetables are becoming ever more demanding. Flavour and appearance determine the choice. The purchasing power of the supermarket makes it difficult for individual farmers to earn a decent living.

Following a merger, the agricultural producers organisation *Groene Hoed* opened up the prospect for farmers to deliver their quality products directly to the customer. A separate brand would need to be created, preferably with its own unique sales channel, for high quality produce to be brought responsibly to the market.

The brand came about. Under the label *MijnBoer.nl* ('MyFarmer.nl') the combined producers delivered their products to supermarket Marqt, restaurant chain La Place and food service company Vitam. The driving force behind these developments was *Marco Duineveld*, for whom this brand meant the translation of his dream into reality: the delivery of fresh quality products by proud farmers to satisfied customers. Duineveld also scaled up the organisation on the producer side, acquiring an environmental hallmark for the entire product range. Ultimately MijnBoer became part of the wholesale group *Sligro/Smeding*, where the product is presented as its fruit and vegetable quality brand. The initiative was evidently so good that the established businesses were unable to get round it and adopted it eagerly. The vision has therefore become a reality that holds out prospects for the future.

Rondeeleggs packed in their unique round, naturally biodegradable coconut-fibre carton (Photo: Mugmedia, Wageningen)

7. RONDEEL

One of the examples repeatedly cited by opponents of the bio-industry concerns the conditions in which chickens are kept. Until a supplier of housing systems for the poultry sector came along who wanted to have a commercial future. As he saw things, this would only be possible by starting from a different position. Not the cheapest henhouse but the one in which the chickens feel most at ease would have the best prospects for the future.

In order to achieve this all kinds of knowledge was of course required. A team was set up consisting of technicians, scientists and people who knew about marketing and brand-building. But perhaps the smartest move of all was the choice of *Ruud Zanders* as a project-leader: a man who, after years of working closely with them, enjoys the confidence of the Animal Protection Foundation. The project also rapidly forged ahead with the elimination of an intermediary: the egg-dealer.

Partly on account of TransForum's input it proved possible to bring a number of players together who managed not only to come up with a henhouse that might have been designed by the chickens themselves but also to gain public appreciation for the eggs that were produced. And all this at a price the consumer was prepared to pay and that would enable the farmer to make a decent living.

The result was a system that went a whole lot further than a henhouse, in which the chickens had all their desires met: night quarters, daytime quarters and outdoor facilities! The necessary steps were also taken to generate enthusiastic backing on the part of both animal rights supporters and the leading supermarket chain in the Netherlands.

Care farms help clients to reintegrate into society

8. LANDZIJDE

The homeless, people who feel lost in the city, drug addicts caught up in the urban drug scene and people suffering from burnout are given a structure, peace and quiet, greenery, space and work on the farm. This helps them get on top of things again and to return to the urban hurly-burly. This too is typical metropolitan agriculture: farmers earn from the services they provide to the city, while this kind of care farming also means that farmers and the typical agricultural landscape are preserved for the city.

Jaap Hoek Spaans is a former teacher and farmer, whose personal drive has been responsible for helping people reintegrate fully into society. He also realised that as a result of the regulations in respect of the landscape and nature conservation, farmers in the Waterland region, a region north of Amsterdam, were finding it increasingly difficult to stay in farming. This exodus was also something he wanted to reverse. This led him to the idea of rolling out professional care farming in the region.

The TransForum project put Jaap in touch with psychiatrists, the Health Council and care and welfare institutions from the Waterland region. This speeded up the professionalisation of the organisation and meant that these bodies now have a place on the Supervisory Board of Landzijde and that joint courses are organised.

The initiative has a high 'cuddliness factor'(meaning that it's really impossible to be against it) and therefore received a lot of political support. The care institutions were positive, while insurance companies contributed ideas. Positive articles appeared in the regional press. As a result, politicians and administrators were glad to be involved, and the idea fitted in with their vision of linking up city and countryside. Insurance companies saw it as a new product in the insurance packages offered to their clients.

Even so, bureaucracy and regulation were the biggest hurdles. Although civil servants were keen to help devise solutions, rules and structures throughout the obstacles. This might for example take the form of zoning plans that permitted agricultural activities in a particular area, while also making it difficult for 'intensive care farming' to be slotted in. Or there might be confusion between all the different budgets from which clients could be funded, and the compartmentalised organisations all of which had to be consulted.

Landzijde is now up and running and enjoys a lot of support. The quality hallmark, certification and supervision are now the important areas of concern.

Cows wandering in the field in the Northern Friesian Woods (Photo: Mugmedia, Wageningen)

9. NORTHERN FRIESIAN WOODS

The inhabitants of the Northern Friesian Woods have a strong sense of cohesion and have traditionally been used to solving their own problems without undue interference from outside. The farmers have their own agricultural nature conservation associations, through which they manage the natural countryside and landscape themselves.

Their region was, however, designated by the 'metropolitan citizens' as a national landscape so as to preserve small-scale landscapes. Now that their incomes have been squeezed, this has set farmers to thinking how they can obtain an income from 'their' landscape. '*Wat smyt it up?*' was the Friesian refrain ('How much does it pay?'). From that point on the goal was clear: the farmers had to earn from their efforts to preserve the landscape and environment.

Politically, Friesland Provincial Executive member *Anita Andriessen* was an important source of support for the farmers. Members of Parliament and the ministers of Agriculture *Veerman* and *Verburg* regarded the area as an experimental location for new policies and accordingly gave it room to experiment. Local aldermen, civil servants, the Dutch Organization for Agriculture and Horticulture (LTO) and

other entrepreneurs provided support and ideas. A group of committed researchers from Wageningen University – some of them of Friesian origin – provided the farmers with a lot of support by conducting research to back the experiments.

All sorts of rules obstructed both the landscape management and the experiments into new forms of income. If for example a branch from a wooded bank was found hanging from barbed wire, the farmer's landscape grant would immediately be cut. Or if the farmers' land contained a pingo (a geological remnant from the ice age), each of the four farmers was required to declare the relevant management activities for a quarter each.

The new product-market combinations have now been designed into business cases. The next step is that of actual investment and implementation. A small group of entrepreneurs is investing a great deal of energy in networking, conferencing and lobbying. By doing so they have gained influence at the national government and have become the most important lobbying partner for the province. The farmers consult with municipalities and other organisations concerning the development of the national landscape, with the farmers' association often taking the lead – a unique arrangement in the Netherlands!

THE NINE PROJECTS IN DETAIL

The next chapters provide relevant information under the same standard headings. This makes it possible for you to selectively look up the information in which you are interested:

1. **The challenge:** describes the entrepreneur's ambition
2. **How did the innovation come about?:** describes the process with which the entrepreneur put the innovation into practice
3. **Key figures**
4. **The added value:** describes the sustainability performances and provides a SWOT sustainability analysis
5. **The value creation model** (for an explanation see below)
6. **From plan to investment:** describes how the plan has been designed in such a way as to lead to an investment
7. **The lessons for the entrepreneur:** lessons from the project, providing an overview of the various roles the entrepreneur must be able to perform for a project to succeed, i.e. the guises he must be able to adopt (see chapter 2)
8. **The current challenges.**

The projects have been clustered according to the three guiding strategies:

1. *'Sustainable Intensification'*: Biopark Terneuzen, Greenport Shanghai, Koe-Landerij and New Mixed Farm
2. *'Sustainable Valorisation'*: Landmarkt, MijnBoer and Rondeel
3. *'Sustainable Diversification'*: Landzijde and Northern Friesian Woods

Notes on the value creation model

A value creation model is provided for each project. The model describes the factors that affect the success of a project or innovation.

The value creation model consists of four interrelated components. According to this model it is vital for distinctive or competitive capacity to be created for agricultural innovations. The distinctiveness of an innovation, or USP (Unique Selling Point), leads ultimately to results. These visible results make it possible to make specific investments that provide the basis for the USP. These investments give rise to competencies that are required for and used in the ongoing creation of distinctiveness. The USP in turn leads to results, thereby ushering in the next value circle.

The value creation model therefore provides a response to questions such as:

- *'What is the USP or distinctiveness of the innovation and what results have arisen from this?'*
- *'What investments have been made?'*
- *'What competencies have emerged as a result and been used in the creation of distinctiveness?'*

SUSTAINABLE INTENSIFICATION

Biopark Terneuzen, Greenport Shanghai, Koe-Landerij, New Mixed Farm

1. BIOPARK TERNEUZEN

Synergy between greenhouse production, chemicals and energy

1.1 The challenge

Society wants businesses to deal with energy, soil, water and other commodities and resources more sparingly. The same applies to reducing all kinds of emissions. In the case of greenhouse horticulture, energy conservation and the reduction in CO_2 are very much a live issues.

New links between various sectors and the physical clustering of various forms of commercial activity make it possible to meet these public demands. In the case of Biopark Terneuzen, this applies in particular to greenhouse production and chemical companies. The unique feature of Biopark Terneuzen is not just this linkage between the agricultural and chemical sector but also the scale on which it is taking place. Biopark Terneuzen is one of the first large-scale industrial agroparks to have come on stream in Europe.

1.2 How did the innovation come about?

A start on the development of 'Valuepark Terneuzen' – a joint venture between Zeeland Seaports and Dow Benelux – was made in 2003. The project seeks to bring together the production and distribution of chemical products in a single location.

Agroproduction parks have formed part of the economic policy of the Province of Zeeland since the year 2000. In 2005 the Province asked the consultancy Van de Bunt to investigate the feasibility of developing agro-industrial clusters in the region. At the same time Yara, a manufacturer of fertiliser, had formulated plans for delivering heat and CO_2 to greenhouse production businesses planning to set up in that area. Van de Bunt discovered that other firms also wanted to invest in the region and saw opportunities for linking up the secondary processes.

Van de Bunt brought together the Province, the Municipality of Terneuzen, Zeeland Seaports and a number of other businesses in order to conduct a number of feasibility studies with the support of TransForum. Detailed commercial and technical feasibility studies were carried out in 2006 – 2007 in order to determine the scope for the development of an agropark and to produce detailed designs. The scope was then widened out from the production and distribution of chemical products to combining these with agricultural products (from greenhouse production) in the one location. Working sessions were held with businesses in order to explore the most promising inter-company linkups. Since the recruitment of greenhouse growers for the Biopark had still to take place, no individual greenhouse growers were involved in the feasibility studies.

In February 2007 *Biopark Terneuzen* received the official go-ahead. Company directors and aldermen jointly raised the Biopark flag. This positioned the Biopark cluster as a brand and made it possible for interested parties to express interest in joining the cluster.

Zeeland Seaports has been an important driver of the Biopark, a role which it continues to play to this day. Zeeland Seaports is the port authority that manages and develops the ports of Flushing and Terneuzen in the interests of regional prosperity. The Province of Zeeland and the Municipality of Terneuzen have a significant voice in the port authority.

Project partners

Bio Glas Terneuzen, Biomassa Unie, Cargill, Municipality of Terneuzen, Heros, Nedalco, Province of Zeeland, Radboud University Nijmegen, Rosendaal Energy, TransForum, VU University of Amsterdam, Wageningen UR (F&BR, LEI and PPO), Yara, Zeeland Seaports and ZLTO.

TransForum project
2006-2007

Under their leadership of the Biopark Terneuzen platform was set up. The latter has since been converted into a foundation with representatives from four public parties and the private sector.

The firm WarmCO$_2$ was established in 2007 for the delivery of heat and CO$_2$ to the greenhouse complex and is now in operation. Bio Glas Terneuzen B.V. is responsible for attracting greenhouse growers to the greenhouse complex.
Over the coming years Bio Glas will be developing a 250 ha greenhouse area, encircled by 60 ha of green space, in three stages.

To begin with, large-scale greenhouse growers favoured having their own combined heat and power plants (CHPs). Had this been allowed this would have ruled out the feasibility of the delivery of industrial CO$_2$ and heat to the greenhouse growers. Zeeland Seaports therefore decided not to accede to this wish. Greenhouse growers wishing to set up in this area are obliged to draw industrial CO$_2$ and heat from WarmCO$_2$.

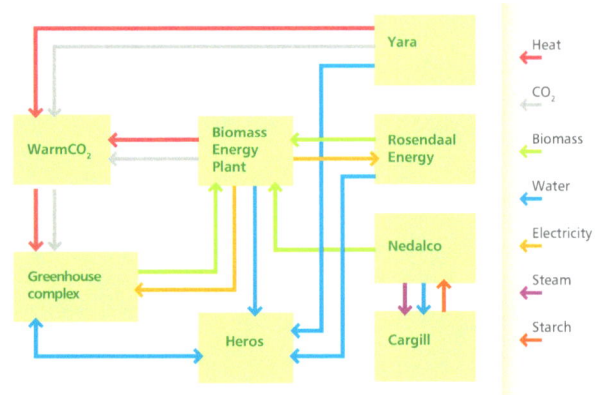

Heat and CO$_2$ from the above streams are now being supplied by Yara to WarmCO$_2$. WarmCO$_2$ in turn supplies heat and CO$_2$ to the greenhouse complex by means of a pipeline several kilometres in length. Of the 125 hectares available in phase 1, 60 hectares have now been sold and 23 hectares are under cultivation: 10 hectares by Tomaholic (tomatoes), 8 hectares by Kwekerij De Westerschelde (sweet peppers) and 5 hectares by Gebroeders van Duijn (aubergines).

Sustainable energy within Glastuinbouw Terneuzen

The sustainable energy concept within Glastuinbouw Terneuzen (Greenhouse Terneuzen) involves the delivery of residual heat and CO$_2$ obtained from Yara, a fertiliser manufacturer located immediately to the north of the greenhouse area. The heat and CO$_2$ released in Yara's production processes would normally be lost. But by capturing these and channelling them via WarmCO$_2$ to the greenhouses in the area, the valuable heat and CO$_2$ are used highly efficiently.
To this end the growers conclude a multi-year contract with WarmCO$_2$ that provides a competitive alternative to a CHP plant. The agreement with WarmCO$_2$ will provide at least 15 years of certainty with regard to energy costs, and savings of at least 80-90% on natural gas consumption. WarmCO$_2$ has constructed its own 5 km pipeline network for the distribution and delivery of the residual heat and CO$_2$.

Source : www.glastuinbouwterneuzen.nl

Biopark Terneuzen

The reduction of CO_2 emissions in glass horticulture is a topical issue. One way forward is to establish new links between businesses in other sectors and glass horticulture so as to reduce CO_2 emissions.

+ The advantages

Fertiliser production

Heat and CO_2 are lost in the fertiliser manufacturing process. By capturing them they can be used as energy for third parties.

WarmCO$_2$ (Storage of heat/CO$_2$)

Heat and CO_2 are delivered to the glass horticulture industry.

Greenhouse complex

Greenhouse horticulture makes use of the heat and CO_2 in order to reduce the impact on the environment.

+ Reduction in the discharge of heat to the Western Scheldt

+ **80%** reduction in carbon footprint

+ **85%** lower use of fossil fuels for horticultural products

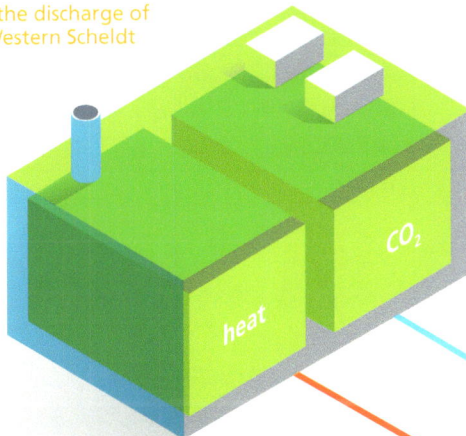

WarmCO$_2$
Maximum capacity:

- sustainable energy: **84 MW**
- greenhouse area: **250 ha**

Area set aside for greenhouses:

total first phase: **125 ha**
of which now in use: **23 ha**

8 ha	**10 ha**	**5 ha**
sweet peppers	tomatoes	aubergines

CO$_2$

heat

Terneuzen Biomass Plant
(under construction/finished 2010)

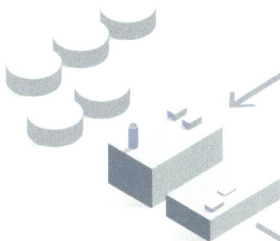

biomass

fermentation capacity:
10,000 MW
23,000 households

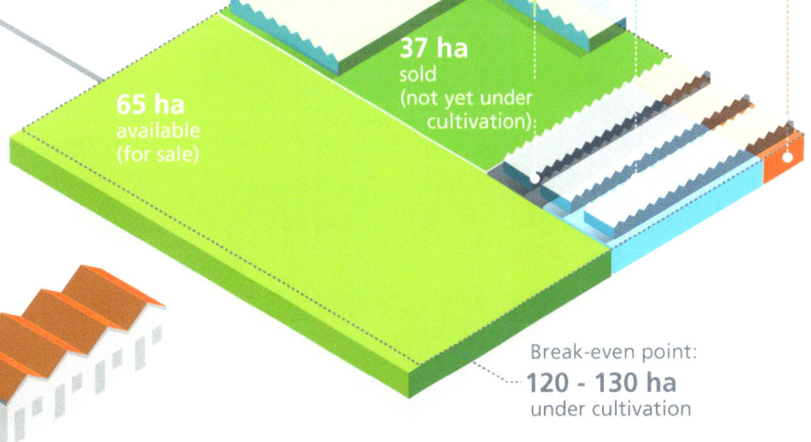

65 ha
available
(for sale)

37 ha
sold
(not yet under cultivation):

Break-even point:
120 - 130 ha
under cultivation

The economic circumstances and resultant reluctance by the banks to lend mean that as in other greenhouse areas, land has not been issued as quickly as expected.

From this it may be seen that a blueprint does not work when it comes to complex innovations. The reality always differs from the plans – and those plans were and remain ambitious. The fact that not all the plans have so far been realised does not detract from the fact that this has already been a considerable achievement.

The current position in relation to the other originally planned linkages is as follows:

- The biomass (fermentation) plant is currently being built at the Heros site. The plant has since been sold by Biomassa Unie to the Lijnco Green Energy. This powerplant processes the biomass obtained from the greenhouse complex. The fermentation plant is expected to come on stream in the fourth quarter of 2010.
- The Rosendaal Energy biodiesel plant has been completed. On account of the rapid turnaround in the market for biodiesel, Rosendaal Energy has gone into receivership and the plant is up for sale. The planned streams from and to Rosendaal Energy have consequently failed to come about.
- The new Nedalco plant has been cancelled. This means that all planned streams to and from Nedalco are not going ahead.
- The exchange of water between the greenhouse complex and Heros has not yet been realised.
- The delivery of heat, CO_2 and power from the fermentation plant to WarmCO2 has not yet materialised.

Further details on the reasons for these alterations may be found later in the chapter.

In contrary to the development of another example of the 'Sustainable Intensification' strategy – New Mixed Farm – there has been little if any public resistance to the development of Biopark Terneuzen. An explanation may be provided in terms of the following differences from New Mixed Farm:

- There is no animal-based agricultural production. This greatly affects the extent to which it is necessary to 'fight' for space in which to produce and operate.
- The Biopark is being constructed in an (agro-) industrial area.
- A deliberate strategy of compensation has been pursued: the nature conservation area de Groene Knoop has been laid out by way of nature compensation.

Cooperation with societal organisations was consequently not such a pressing requirement in this project. The collaboration among various chemical companies and between greenhouse growers and chemical companies has been an important achievement. It is this cluster of connections on this scale between the chemical industry and the agricultural sector that makes this project so special.

1.3 Key figures

Position as of August 2010

Greenhouse cluster

- 240 hectares of net greenhouses of which 60 hectares have been sold and 23 are under cultivation: 10 ha by Tomaholic (tomatoes), 8 ha by Kwekerij De Westerschelde (sweet peppers) and 5 ha by Gebroeders van Duijn (aubergines).
- WarmCO2 with a maximum capacity of 84 megawatts of sustainable energy (132 megawatts of total heat during the winter peak), sufficient for 168 hectares of greenhouses.

Biomass cluster

- Biomass (fermentation) plant is under construction, with a capacity of 10 tonnes Megawatt. This is equivalent to the energy consumption of 23,000 households or 100,000 persons: a fifth of total household consumption in Zeeland.

2006-2007	Various technical and commercial feasibility studies
February 2007	Official go-ahead for Biopark Terneuzen; hoisting of the Biopark flag
2007	Establishment of WarmCO$_2$ and Bio Glas Terneuzen B.V.
Late 2007	Acquisition of land for greenhouse area by Zeeland Seaports completed
Late 2007/ early 2008	Commencement of development of nature conservation area De Groene Knoop (nature compensation)
November 2008	Definitive approval of zoning plan for Kanaalzone Greenhouse Area in Terneuzen
April 2009	Commencement of construction of Gebroeders Van Duijn greenhouse (aubergines)
Summer 2009	Commencement of construction of Kwekerij Westerschelde (sweet peppers) and construction of Tomaholic greenhouse (tomatoes)
October 2009	WarmCO$_2$ receives € 25 million green grant
November 2009	First delivery of residual heat to greenhouse growers by WarmCO$_2$
Late 2009	First aubergines, sweet peppers and tomato plants planted
January 2010	Construction of phase 2 of De Groene Knoop nature conservation area gets underway (nature compensation)
Spring 2010	First harvest of greenhouse vegetables
2010	Construction of the biomass plant; commissioning expected in Q4 2010

1.4 The added value of Biopark Terneuzen

The advantages of the greenhouse complex at Biopark Terneuzen

- The clustering of agricultural and chemical firms enables them to make use of each other's waste and residual flows, generating sustainability benefits.
- The area offers expansion possibilities for greenhouse growers.
- The area offers good logistical connections with Belgian auction houses.
- Greenhouse growers gain access to pure CO_2: vital for good production yields.
- The recycling of industrial residual heat and CO_2 mean that the vegetables grown here have a lower environmental footprint.
- The Horticultural Information and Education Centre ensures a supply well-trained workers.

The sustainability performances

A brief summary of the most important results of the greenhouse element in Biopark Terneuzen is provided below, divided into profit, people and planet aspects and relevant underlying sustainability aspects.

People

- The environmental quality score is positive. Yara is no longer discharging surplus heat into the Western Scheldt river basin since this heat is now used for the greenhouse complex.

Planet

- 80% lower carbon footprint of the cultivated products (CF). In the case of aubergines this is 0.4 kg CO_2 per kg of aubergines as compared with over 1.6 kg CO_2 per kg in the case of aubergine cultivation for which heat is provided solely by boiler. Compared with aubergine cultivation working with both CHP and a boiler, the carbon footprint of aubergines grown in the Biopark is 44% lower. This substantial reduction in the carbon footprint is likely to apply to all greenhouse vegetables in this Biopark cluster.

- 80-90% lower consumption of fossil energy (especially natural gas).

Profit

- Potential commercialisation of the lower environmental impact of the greenhouse vegetables grown in Biopark Terneuzen a) via a higher price due to the lower environmental impact or b) greater ease of concluding delivery contracts due to the sustainability gains.

SWOT analysis of the sustainability performances

Strengths

- Big reduction in greenhouse gas emissions and much less use of fossil energy for the greenhouse products grown in Biopark Terneuzen.
- Improved environmental quality due to the reduced heat discharge to the Western Scheldt.

Weaknesses

- The investments in the new infrastructure for the delivery of waste flows and by-products are costly. The investment cost for the streams of heat and CO_2 from Yara to the greenhouse area was around 80 million euros.

Opportunities

- Establishment of new greenhouse enterprises. The growing demand for environmentally-friendly cultivated products creates opportunities for value creation.
- The development of emission rights.

Threats/risks

- No new greenhouses or bankruptcy of already established greenhouse enterprises.
- WarmCO$_2$ or Yara go out of business.

1.5 The value creation model

Notes on the value creation model

Value can potentially be created in two different ways in Biopark Terneuzen. First of all value can be created in respect of the greenhouse vegetables. Two of the three greenhouse vegetable growers intend to market greenhouse vegetables from the Biopark on the basis of their lower environmental impact. Demand for climate-neutral vegetables is, for example, growing in the UK market. This provides opportunities for these entrepreneurs. They are not yet embarking on this marketing strategy as the certification process is demanding and calls for considerable investment. In order to do so and market climate-neutral vegetables, more enterprises will be required.

The second potential way in which value could be created is the application of biomass for the generation of sustainable electricity, which is then sold. This biomass (fermentation) plant is under construction now that a sustainable energy promotion (SDE) subsidy has been granted for a period of eight years. At the end of that period the fixed costs will have been written off and it will be possible to operate on the basis of variable costs. A favourable ROI (Return On Investment) is therefore anticipated for the next eight years and beyond.

The fermentation plant has a capacity of 10 tonnes Megawatt, equivalent to the energy consumption of 23,000 households or 100,000 people. This would enable the fermentation plant to supply a fifth of the total household energy consumption in the Province of Zeeland.

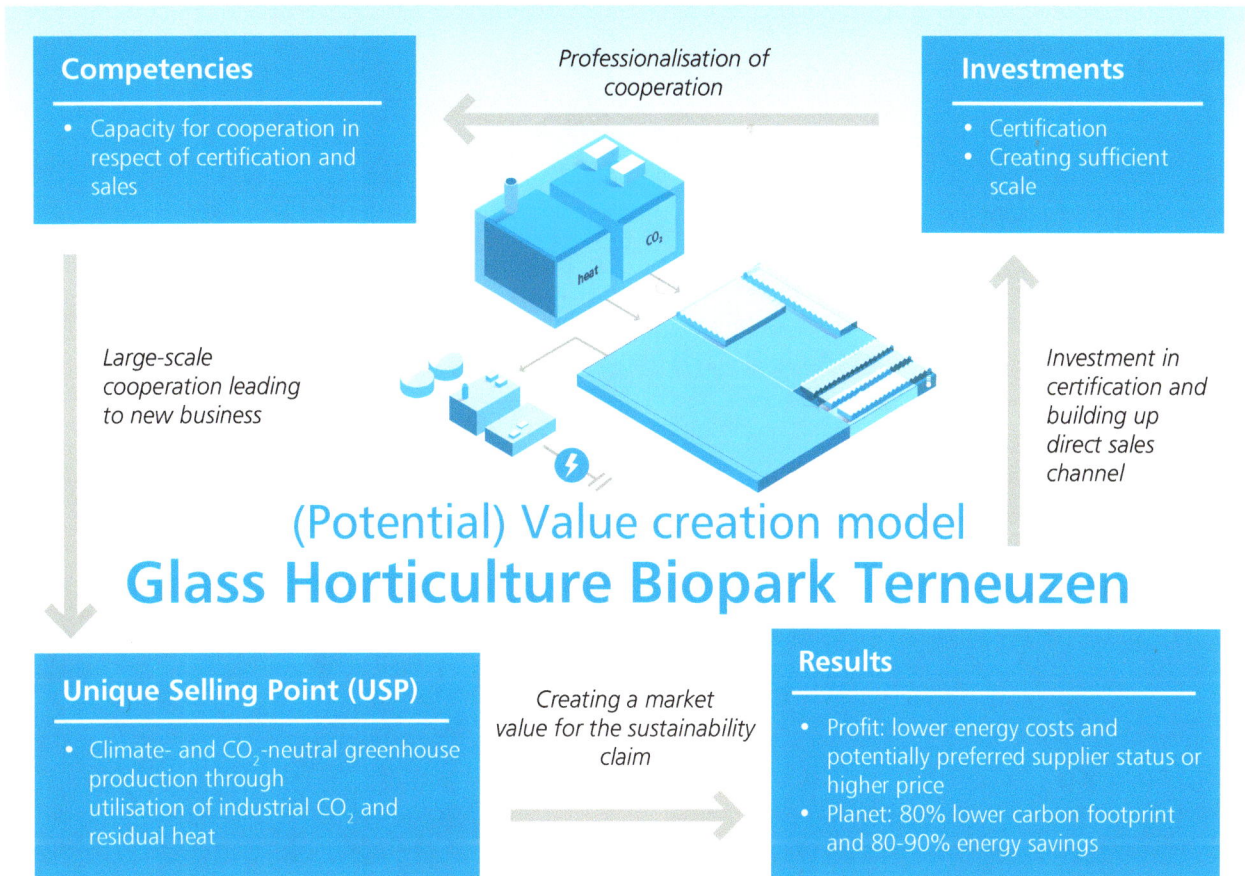

Competencies

• Capacity for cooperation in respect of certification and sales

Professionalisation of cooperation

Investments

• Certification
• Creating sufficient scale

Large-scale cooperation leading to new business

Investment in certification and building up direct sales channel

(Potential) Value creation model
Glass Horticulture Biopark Terneuzen

Unique Selling Point (USP)

• Climate- and CO_2-neutral greenhouse production through utilisation of industrial CO_2 and residual heat

Creating a market value for the sustainability claim

Results

• Profit: lower energy costs and potentially preferred supplier status or higher price
• Planet: 80% lower carbon footprint and 80-90% energy savings

---

Since the plant will be commissioned at the end of 2010, we shall concentrate here on the elaboration of the 3P added value for the greenhouse complex within Biopark Terneuzen.

The creation of value by the greenhouses at Biopark Terneuzen is shown in a model above:

It would be very helpful if the climate-neutral and CO_2-neutral method of production could also be recouped by charging a higher price for the glasshouse vegetables. The value creation model for the Biopark greenhouse vegetables could turn out as shown above if the growers manage to exploit the sustainability benefits in the market.

This would require production on a bigger scale. Product recognition in the market would be facilitated by a hallmark, for example the Dutch environmental hallmark *Milieukeur* or the HIER logo obtainable from regional environmental federations. This requires certification, which is not easy and calls for a joint investment by greenhouse growers in Biopark Terneuzen. Larger-scale production is also required as the placement of a recognisable product in the market is easier by means of direct sales. To be interesting to customers this will need to be on a sufficient scale and investments will have to be made in relationship management.

At present such products are being sold at auction and are still relatively anonymous. If the greenhouse growers wish to go in this direction, cooperation with colleagues from the greenhouse cluster at Biopark Terneuzen could offer economies of scale and the number of greenhouse firms there will need to grow.

1.6 From plan to investment

What is notable about this project is that no business plan was developed by the entrepreneurs and government authorities. Following the various feasibility studies, individual firms set up on the basis of their own business plans. The initiative to make connections (or 'smart links' as they say in this project) came from local and regional government. The links ultimately developed among the enterprises are located at the level of the secondary, supporting processes. The linkage of secondary processes leads on the one hand to sustainability benefits and, on the other, to additional costs and risks and to some extent duplicate investments in necessary backup systems.

Investment and financing of the Yara - WarmCO$_2$ – greenhouse cluster

WarmCO$_2$ is a project of Zeeland Seaports, Yara and Visser & Smit Hanab. A total initial investment of 80 million euros was made in the development of WarmCO$_2$.

The bank ABN AMRO provided WarmCO$_2$ with 25 million euros in 'green finance' under the Green Projects Regulations, a joint scheme of the Ministries of Finance, Agriculture, Nature and Food Quality (LNV) and Housing, Spatial Planning and the Environment (VROM), which offers tax benefits to 'green' savers and investors. This enables banks to offer loans at a lower rate of interest. WarmCO$_2$ is the first large-scale sustainable energy project to have received a green loan. The total financing of WarmCO$_2$ amounts to 80 million euro. 15 million euros were invested using subsidies and the remainder with loans and equity capital. Yara put in 1 million euros and also invested heavily in staff training and the transportation of heat from the plant at Sluiskil to WarmCO$_2$.

WarmCO$_2$ has a maximum capacity of 84 megawatts of sustainable energy (132 megawatts of total heat during the

*Aubergine grower Rob van Duijn buys heat and CO$_2$ from WarmCO$_2$
(Photo: Mugmedia, Wageningen)*

winter peak), sufficient for 168 hectares of greenhouses. If this capacity is 100% utilised, the ROI is 7%. At present 23 hectares are in use. The breakeven point for WarmCO$_2$ is between 120 and 130 hectares.

For the greenhouse growers the initial investment in their business amounts to 8-10 million euros for 10 hectares. In addition they need to invest in a diesel generator to cover any breakdowns in the supply of heat by WarmCO$_2$. This is not in fact an additional investment as a CHP plant also calls for a backup system. The initial costs in this case therefore amount to 1 million euros per hectare.

The ROI of the greenhouse growers in Biopark Terneuzen depends on the market prices of the greenhouse vegetables they produce. In this case it is interesting to examine the share of energy costs in the total costs. According to the greenhouse growers, this amounts to approximately 13-18% of the total cost. CO$_2$ accounts for 6-8% and heat for 7-10% of the total cost. The greenhouse growers have a contract with WarmCO$_2$ for a fixed energy price for 20 years.

At the present time these heating costs are comparable with those of alternative systems such as CHP. Much depends on the extent to which a greenhouse growers manages to conclude an attractive deal with an energy supplier for the offtake of gas and export of electricity to the grid. If fossil energy prices rise this will give the greenhouse growers in Biopark Terneuzen a competitive advantage.

According to the greenhouse growers themselves, their reasons for setting up in Biopark Terneuzen are not primarily related to the link with WarmCO$_2$. The most important reason is that they saw the opportunity to expand their business, something which is often not possible in other greenhouse areas. For others the good logistics are a factor. This applies especially if the vegetables are sold at auction in Mechelen or Venlo. The greenhouse growers also expected the availability of labour in this region to be an advantage. That has not been particularly true. Greenhouse production is comparatively new in this region, and 'unknown means unloved'. The Information and Training Centre for horticulture in Terneuzen

(www.werkenindekas.nl) is designed to bring about a change. Attracting students and potential employers has been more difficult than expected as unemployment in this region is relatively low.

1.7 The lessons for the entrepreneur

The entrepreneur as project developer

The plans were not subjected to a sufficient reality check during the feasibility study stage of this project. Entrepreneurs were poorly involved in this phase. Not all the plans proved realistic at the investment stage. The delivery of heat, CO$_2$ and power from the fermentation plant to WarmCO$_2$ was for example devised on the drawing board, but has failed to materialise. The involvement of entrepreneurs/investors right from the planning stage will prevent unrealistic plans from being drawn up.

The entrepreneur as spider in the web

The initiative to establish links came from the local and regional government. As the representative of these governments, Zeeland Seaports took on the role of standard-bearer and pioneer. The company invested a lot of time in hooking other parties up to the Biopark Terneuzen concept.

The Van de Bunt consultancy took the initiative to bring firms interested in the area together around the theme of closing loops and agroproduction parks. This put flesh on the bones of the concept of industrial ecology for the first time in this area.

The market launch of Biopark Terneuzen was given symbolic force by the hoisting of the Biopark flag in February 2007, thereby sending the story out into the world as a brand. Zeeland Seaports was also responsible for orchestrating the substantive development of the agropark. A vital factor was that Zeeland Seaports decided not to permit greenhouse growers to run their own CHP plants. This was a key element in safeguarding the concept in relation to the greenhouse cluster of Biopark Terneuzen, and ensured that the linkup of Yara, WarmCO$_2$ and the greenhouse cluster became viable. Right from the start strong local and regional support was provided for Biopark Terneuzen. The initiative was supported by both the Municipality of Terneuzen and the Province of Zeeland and Zeeland Seaports. They continued to do so throughout the process. In addition, there was little if any local and/or public resistance to the plans.

The entrepreneur as winner

Changes to the primary process can sometimes be needed for the waste streams of other businesses to be turned to commercial account. This can lead to undesirable commercial risks and the need to put in place extra backup systems, thereby increasing the cost and investment required. In the case of Yara for example the alterations to the heat exchange systems had a major impact on the reliability of the primary process and hence on the continuity of the business. In order to reduce the risks, backup systems were installed, with a consequent increase in the total investment cost.

The greenhouse growers invested in a backup system in the form of a diesel generator. In fact all the businesses took measures to make sure that they could operate on a stand-alone basis. The reduction of dependency is a standard element in commercial risk management. This may be at variance with the links outlined on the drawing board from a sustainability viewpoint.

This case indicates that it can be risky for entrepreneurs to set up a business in government-oriented markets directed towards sustainability. In recent years, activities aimed at a biobased economy have been hyped up and subsidised on a large scale. Does this mean that entrepreneurs are in tune with the times, or in fact the very opposite? In an optimistic atmosphere of this kind, there is a tendency to neglect to ask the right questions about the feasibility, risks and robustness

of the business model should government support be withdrawn or the actual production capacity exceed market demand.

As this case study shows, this can be risky. The lowering by the government of the biofuel additive stipulation redefined the size of the market required by Dutch Biodiesel plants, including that of Rosendaal Energy. Furthermore, the support provided by various enthusiastic governments with readily available subsidies encouraged the uncoordinated construction of more plants and greater capacity than the market could cope with. In addition the biodiesel market became heavily dominated by cheap US imports. This combination of circumstances had the result that all seven biodiesel plants in the Netherlands have (at least for now) been closed down. The Rosendaal Energy biodiesel plant was in production for six months, but the company has gone bankrupt and Zeeland Seaports is looking for a new buyer for the plant.

WarmCO$_2$ decided at a certain point to accept all the risks involved with the construction of the infrastructure. On the one hand this decision ensured that the low-energy and climate-neutral greenhouse area could get off the drawing board. On the other hand, the fact that no other players were prepared to share the risk may say something about the design of this linkup between industry and glass horticulture. An alternative choice might have been to examine the design more critically and where necessary make modifications.

What is notable is that no effort appears to have been made at agropark or park management level to explore other forms of added value – such as new products or services – arising from the synergy between the various elements. Instead, the concentration was on linking up secondary processes. In other words, the focus was on the reduction of costs rather than the creation of new value, and on *'Sustainable Intensification'* rather than on *'Sustainable Diversification'*. There are therefore more opportunities in this area for the development of new, sustainable commercial activities.

1.8 The present challenges

The greenhouse/ WarmCO$_2$ /Yara cluster faces at least two challenges in the near future.

The first is to continue developing the greenhouse area in the coming years. The breakeven point for WarmCO$_2$ is an area of between 120 and 130 hectares under glass. At present 23 hectares are under glass. This means that a further 100 hectares under glass will need to be added within the foreseeable future. In addition it would be very helpful if the climate-neutral and CO$_2$-neutral method of production could be recouped by charging a higher price for the greenhouse vegetables. This would require a bigger scale of production, as well as cooperation.

Greenport Shanghai and New Mixed Farm are agroparks that are still at the planning stage or on the verge of implementation. Biopark Terneuzen by contrast is an established, operating agropark: an accomplishment within just three years of which to be proud.

Critical voices will of course wish to emphasise that at least half the ambitions remain to be realised. That is the flipside of large-scale ambitions. As far as we are concerned, however, the glass here is clearly half full, not half empty. The enterprises in Biopark Terneuzen have opted for a step-by-step approach, meaning that the ambitions cannot all be realised in one fell swoop.

A POWERFUL PIONEER IS CRUCIAL

Commercial manager of Zeeland Seaports, Peter Geertse
(Photo: Mugmedia, Wageningen)

2. GREENPORT SHANGHAI

model for mass agglomerations

2.1 The challenge

The world is becoming urbanised: more and more people are living in cities. The rapid growth of the big cities is increasingly placing a double squeeze on agriculture. On the one hand the growing urban population has an ever-growing need for safe, high-quality food. On the other hand, the urban growth competes with agriculture for space, with agriculture often coming off second-best. In China no more than 13% of the land area is sufficiently fertile for agriculture. It is precisely in these scarce fertile regions that the urban growth is concentrated. In this situation reliable food chains delivering fresh products become strategically important.

In the case of a rapidly growing city as Shanghai, the question arises as how to meet the demand for fresh agricultural products for the growing population when land, water, energy and nutrients are becoming ever scarcer. In addition, the rapidly growing middle class in Shanghai has taken fright as a result of recent food scandals and become ever more critical of food safety and quality, setting a standard that cannot in fact be achieved by traditional Chinese agriculture and its logistical system.

The challenge in this project was accordingly fourfold:

1. Increasing the level of production of fresh food in the immediate vicinity of Shanghai in order to cope more effectively with the rapidly growing middle class of consumers.
2. Improvement of the agro-logistics around Shanghai so as to guarantee food safety and food quality more effectively.
3. Realisation of agroproduction with optimal water, nutrients and energy efficiency.
4. Turn agriculture into an engine for sustainable urban development.

The Chinese partners regarded the Dutch agroparks concept as a possible solution for these challenges. An agropark is an innovation in the field of agroproduction, processing and logistics. It covers aspects as production, processing, trade, demonstration, R&D, capacity building and social functions, supplies the products throughout the year as efficiently as possible, and is partly independent of seasonal factors and the soil. Greenport Shanghai is a showcase of an agropark, where modern, secular agriculture is practised.

Greenport Shanghai is an agropark consisting of various large-scale agroproduction units, a food processing park, a logistics centre and a research and training centre. By means of a Central Processing Unit (CPU), the residual and by-products of the one production unit are used as input for the other and sustainable energy is generated. The integration and linkage of various kinds of agro-activity in the agropark offer Greenport Shanghai opportunities for a huge sustainability drive, in which the whole is greater than the sum of the parts. Greenport Shanghai therefore fitted in perfectly with the slogan of the World Expo 2010, 'Better City, Better Life'.

For the Netherlands this project offered a large-scale experimental area for showing how large-scale production can make huge sustainability gains. This has opened up a perspective in which Dutch knowledge and entrepreneurship in this field can be commercialised in the world market, thereby enabling the Netherlands to play an international pioneering role.

Project partners
Evelop, Municipality of Venlo, Greenport Nederland, Grontmij, HeadVenture, Holthuis International Lawyers, KnowHouse, Merapeak, Province of Limburg, Shanghai Agricultural Commission, Shanghai Development and Reform Commission, Shanghai Industrial Investment Cooperation (SIIC), Wageningen UR (Alterra and F&BR), Waste Management Middle East and TransForum.

TransForum project
2006-2009

2.2 How did the innovation come about?

At the end of 2006, KnowHouse, Shanghai partner SIIC, TransForum and Wageningen UR set up a project group to jointly design a masterplan under the name 'Greenport Shanghai'. The masterplan contains four scenarios for the spatial and technical design of the agropark, consisting of a production park, a processing park, a trade centre and a demonstration park for large-scale sustainable agro-activities relating to food supply.

The beating heart of the park is the Central Processing Unit. Via a waste-to-energy cycle, the biofermentation plant converts the waste in the form of biomass to new products such as energy, heat, CO_2 and water. These new products then serve as input for the agroproduction and processing firms in the park. The park management organisation offers these services to all the agricultural entrepreneurs in the park in return for payment.

Greenport Shanghai is ambitious in nature and the realisation of the innovative projects consequently involves a number of challenges:

1. Technical integration of large-scale agroproduction, processing and logistics in order to generate synergy benefits (hardware issue: 'what techniques do we require?').
2. Combination of Chinese eco-city ambitions with Dutch knowledge, experience and entrepreneurship into a new development approach (software issue: 'what knowledge do we require?').
3. Setting up a park management organisation to oversee the sustainability performances of the park and offering various services in this regard (orgware issue: 'what organisation do we require?').

In this regard is important is to be aware that Greenport Shanghai involves a business model at two levels:
- Agricultural entrepreneurs have lower costs due to the cooperation in the agropark and the higher water, nutrients and energy efficiency and higher yields from the concentrated and efficient production, plus opportunities of a higher margin through quality guarantees and branding. They pay the park management a fee for the services.
- The park management manages the property and the infrastructure of the agropark. The infrastructure consists not just of basic infrastructure such as buildings and roads but also 'circular' infrastructure, including the installations for water, nutrients and energy management. In the case of the real estate, the park management receives rental, while for the services the entrepreneurs in the park pay a fee. The CPU is dimensioned in such a way that it is not exclusively dependent on biomass flows from the park but is also capable of processing urban waste. This generates additional income, plus surplus energy, which can be sold to the electricity grid.

Below we examine in turn the technical innovation (hardware issue), the social innovation (software issue) and the organisational innovation (orgware issue).

Technical innovation: integration of production, processing and logistics

Greenport Shanghai not only brings various businesses together in the one place but is characterised by the synergy generated among those enterprises. There is both vertical integration (within the chain) and horizontal integration (with other sectors). This integration drive calls for technical linkages and a highly sophisticated spatial design.
The technical linkages in Greenport Shanghai are provided primarily by the Central Processing Unit or CPU. In the CPU, the biofermentation plant converts the waste in the form of biomass to new products such as energy, heat, CO_2 and water via a waste-to-energy cycle. The CPU therefore plays a vital role in the management of nutrients, water and energy in the park. The CPU enables the residual and by-products of the one firm to be used as input by another. Every effort is also made to promote the production of other forms of renewable energy, for example by the use of solar panels and wind energy. In a spatial sense, the configuration of the businesses in the agropark is designed to minimise transport movements. Fertiliser is transported by underground pipe, as are heat, water and CO_2. Harmonisation with the landscape also forms an important element in the spatial design of Greenport Shanghai. By making use of green roofs isolation is also greatly improved. Natural water filter systems, such

2006   During a visit to the WAZ-Holland Park project in Changzhou in spring 2006 an unscheduled meeting took place between representatives of Alterra, KnowHouse and TransForum and a representative of the Shanghai Industrial Investment Corporation, who was interested in the agropark concept. In September 2006 the project proposal for Greenport Shanghai Agropark was approved. In autumn 2006 a mission visited Shanghai with local politicians and entrepreneurs from the Venlo region (consisting of KnowHouse shareholders). The participants saw opportunities for marketing Dutch knowledge and entrepreneurship, such as that from the New Mixed Farm project.

2007   Various design sessions involving participation by Chinese and Dutch knowledge institutes and the agro-industry with a view to developing the masterplan. Summer 2007 completion of masterplan and presentation in China. Masterplan officially approved by SIIC. In October 2007 signature of Memorandum of Understanding for the development of Greenport Shanghai during Seminar on Metropolitan Agriculture in Beijing, presided over by the Dutch and Chinese ministers of agriculture. Appointment of Dutch steering group to help resolve the jockeying for position among the Dutch players.

2008   Limburg agro-entrepreneurs drop out when it becomes clear that the Chinese partner SIIC is unprepared to pay for knowledge, while at the same time expecting Dutch entrepreneurs to undertake the risk-bearing investment. Alterra submits a quotation to SIIC for further consultancy by way of follow-up to the masterplan. NL players make a number of visits to Shanghai in order to clarify expectations, roles and approach. This results in an impasse in summer 2008. Deadlock appears to have been overcome by an initiative on the part of the Province of South Holland and Greenports Nederland for a visit to Shanghai in September 2008. They see opportunities for a leading role for Greenports Nederland and to this end sign a new MoU in Shanghai. Alterra, KnowHouse and TransForum are sidelined by Greenports Nederland. This initiative too soon founders. At the request of the Steering Group TransForum lays the basis for a developmental consortium while KnowHouse makes a renewed effort to interest a group of Dutch agro-entrepreneurs (broader this time than Limburg) in participating in Greenport Shanghai. In December 2008 TransForum, together with Holthuis International Lawyers, submits the business pre-conditions to the SIIC, on the basis of which the Dutch developmental consortium is prepared to set up a joint venture with SIIC for the development and management of Greenport Shanghai.

2009   SIIC indicates that it is not (yet) able to comply with all the business preconditions laid down. A Dutch developmental consortium offers to undertake the first stage of the further plan development on a consultancy basis and to finance 50% of the costs incurred if this is followed by joint-venture. No official reply to this offer is received from Shanghai, although reports are received of changes at the top of SIIC arising from a corruption scandal in Shanghai. Being identified with the president, the project appears set to become the sacrificial victim. On the Dutch side Econcern is wiped out in the economic crisis and the consortium consequently collapses. The project is formally terminated. TransForum does however continue to invest in the development of a generic business plan for the commercial substantiation of the added value offered by this concept. There is clear spin-off from the project in the form of consultancy contracts awarded to Alterra, for example to design agroparks in India, and the initiative for a 'Greenport Holland International'.

as helophyte filters, are also used. This involves not just respecting but also making best use of the natural gradients on the site.

The integration and linkage of firms provide the essential strength of the agropark, but can also be seen as a risk as this assumes a high measure of interdependence. In theory an optimally functioning CPU should have a decisive bearing on the configuration and scale of agricultural production in the park. In the practical world of farming this is of course unrealistic.

Where the masterplan described optimal scenarios, it was soon decided in the implementation phase of Greenport Shanghai that the CPU should not just be measured in terms of the waste flows in the agropark but should also be established as an independent firm with input from outside the agropark as well. The CPU is, accordingly, capable of guaranteeing the offtake of waste and the delivery of energy and other products, especially to the enterprises within the agropark.

The agropark is therefore being realised with the use of existing, proven techniques of individual participants. In practice the agropark consists of an amalgamation of chain activities that generate added value for all concerned. Particularly on account of the reduction in logistical movements brought about by the CPU and the physical clustering of the businesses, major gains have been made in terms of cost and food quality.

Social innovation: new knowledge and new entrepreneurship

Large-scale developments in China are still chiefly directed on a top-down basis. The tradition of 'blueprint planning' remains firmly in place in China. This is however at variance with the development-oriented, participatory approach needed for the realisation of an agropark. This calls for active involvement in the park design by the ultimate entrepreneurs in the park.
In drawing up the masterplan use was made of both scientific knowledge and the experiential knowledge of innovative Dutch entrepreneurs. The combination of these types of know-how has provided a bedrock of knowledge for the Greenport Shanghai plans.
To avoid getting caught up in the pitfall of a blueprint, four possible scenarios were elaborated in the masterplan, indicating specifically that the ultimate design of the agropark depends on which entrepreneurs ultimately want to produce there. Even so, the Chinese partners continued to display a marked preference for selecting and then implementing one particular scenario. Cultural management has therefore been a crucial element in this project.

In retrospect it is evident that the Chinese administrative requirement for a masterplan initially held back the development of a business plan. Dutch agricultural entrepreneurs were primarily involved as suppliers of know-how, i.e. as consultants, rather than as risk and revenue sharing partner (RRSP). The result was that it was primarily scientists who led the formulation of the masterplan and that too little attention was paid to the commercial elaboration of the ideas. The sting in the tail came when the Chinese partners asked the Dutch parties after the completion of the masterplan to help them implement it. At the time none of the agricultural entrepreneurs was either willing or able to take part on a risk-sharing basis in the development of Greenport Shanghai, as the design to emerge from the masterplan was not based on a solid business plan. The assumption that Chinese partners would pay for Dutch know-how therefore proved unfounded.

What the Chinese wanted was 'packaged knowledge' in the form of joint ventures between Dutch and Chinese firms which, with a combination of know-how, entrepreneurship and financial resources, would invest in the development of Greenport Shanghai by jointly setting up new forms of economic activity within the park.

In the process of setting up Greenport Shanghai, a set of design requirements was generated for a new development model that does justice to the need to utilise scientific knowledge while also being business-driven. The starting point for this model is a linked development at both park management level and at the level of the agricultural entrepreneurs. This transcends the traditional developer/customer relationship normally encountered in business park development, and is sometimes termed 'project development plus'.

Organisational innovation: more than the sum of the parts

Only after two years it became apparent that somebody had to be made explicitly responsible for achieving more than just the sum of the parts in the project. This key question did not come up in the project until late in the day. On account of the scale and complexity of Greenport Shanghai the question is a crucial one. During the course of the project a number of options were considered, each with its own pros and cons:

Consortium of Limburg agricultural entrepreneurs

- Advantage: business-oriented approach based on the agro-chain, financial support from the Province of Limburg.
- Disadvantage: project too big and complex to be tackled bottom-up. Requires financial investments that are not consistent with the SME-nature of most Dutch agro-firms. Little knowledge of project-development.
- For some family businesses the distance and cultural differences are an important barrier.

Wageningen University

- Advantage: has in-house expertise and sees an opportunity for marketing the agropark concept worldwide.
- Disadvantage: is generally speaking more knowledge-oriented than business-oriented. Unable to contribute resources of its own and is therefore essentially a consultant for a client. Is not therefore in a position to oversee a concept other than by guiding or coaching the client. That will go only so far as the client permits.

Shanghai Industrial Investment Corporation

- Advantage: obtained development rights for Greenport Shanghai from the Chinese government and is in a position to make a financial investment.
- Disadvantage: is a traditional developer, is not familiar with agroparks, has a short-term interest in attracting foreign investments rather than a long-term interest

in the sustainability performance of the agropark. SIIC earnings model is based on the rise in value of the land, not in adding value to production in the agropark.

Chinese government - Dutch government

- Advantage: fits in well with the Chinese model of strong government direction, consistent with need of Dutch regional and local government to score with this project.
- Disadvantage: frustrates entrepreneurship, the political agenda is determined by the issues of the day, subsidy policy of the Dutch government is aimed primarily at knowledge development rather than commercial development.

The conclusion is therefore that there is as yet no one organisation that is capable of facilitating the development of agroparks such as Greenport Shanghai. An effort was therefore made in this project to set up a new Dutch business consortium aimed at the development and management of agroparks. The consortium initially consisted of Evelop – the project development arm of Econcern, which specialises in sustainable building and sustainable energy infrastructure – and Waste Management Middle East, a company that specialises in waste-to-energy-installations. With TransForum acting as the linking pin vis-a-vis the Dutch agro-business community and the agro-knowledge world, and also as guardian of the concept as a whole, the consortium wanted to take responsibility for the development and management of the real estate and infrastructure in Greenport Shanghai. This was on condition that a park management organisation would be set up in collaboration with SIIC in order to provide the necessary organisational structure for the 'project developer plus' model.

Greenport Shanghai

Greenport Shanghai is a system innovation for agroproduction, processing and logistics. The agropark creates the possibility of closing cycles and reducing transport and more efficient land use.

15-20% annual increase in the demand for agrofood

13% of Chinese land is suitable for agriculture

⊕ The advantages

······ Shanghai

Declining share of small-scale farmers in primary production

Agroproduction Park
Hen housing and pigsties

Mushroom cultivation, outdoor arable farming and greenhouses

Yield

80%   **30%**

2006   2020

400 ha

38 ha

50 m.
eggs

240,000 tonnes
of pig meat

60,000 tonnes
of chicken meat

24,000 tonnes
vegetables

57,000 tonnes
of mushrooms

⊕ greater employment in the long-term with better working conditions

⊕ responding to the needs of the rapidly growing middle class of consumers

⊕ efficient use of nutrients, water and energy

⊕ better animal welfare: unusual for China

⊕ guaranteeing food quality

⊕ energy production from (urban) waste

⊕ reduction in transport

1 m.
meat pigs

0.12 m.
sows

2 m.
laying hens

5 m.
broilers

CPU output:

460,000
tonnes of waste
→ **443 m. kWh**
of electricity

1.7 m.
tonnes of biomass
→ **401 m. kWh**
of electricity

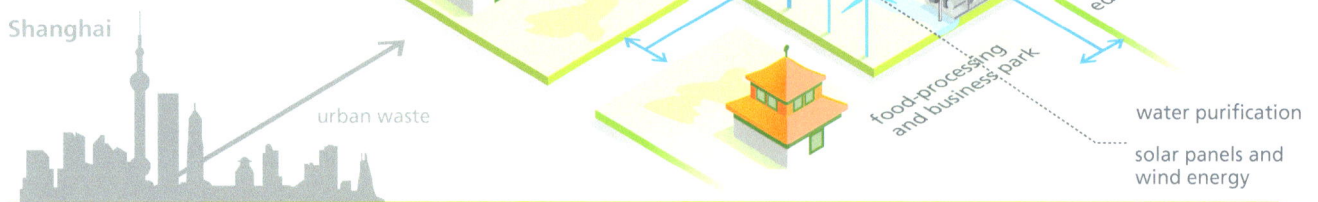

agroproduction park

park management

Central Processing Unit

food-processing and business park

research, training and education centre

Shanghai

urban waste

water purification

solar panels and wind energy

2.3 Key figures

- The Chinese agrosector is huge and constantly changing. Due to the continuous rise in prosperity, the demand for agro-food is rising by 15-20% a year.
- The number of consumers in the middle to top income brackets is rising rapidly (from 455 million in 2006 to 1000 million in 2020). They have different consumption patterns, with a marked need for quality, food safety and food security.
- This results in an increase in the supermarket share from 15% (2006) to 40% in 2020. With regard to the primary production, the share of small-scale farmers will decline from 80% in 2006 to 30% in 2020.
- Just 13% of China's land area is suitable for cultivation.
- Due to the urbanisation, industrialisation and desertification, the availability of agricultural land is under ever growing pressure.
- Partly on account of the smallness of scale of production, the distribution suffers from inefficient logistics.
- Just 30% of consumption in China is processed by processing industry, compared with 80% in the West.
- The market for processed products is, however, growing explosively, especially in urban areas.
- Chinese consumers attach particular importance to safe food – particularly after the food scandals in 2008 – and it is also the intention to export safer, processed products. Ever more stringent demands are being placed on hygiene and safe and healthy food.
- The Chinese government wishes the country to be self-sufficient in food, especially rice and cereals, and is aware of the acute risk of shortages with the reduction in the area of agricultural land. In this regard the government is playing a pivotal role in expanding the number of modern agricultural and horticultural enterprises.
- The province of Shanghai is therefore developing the Dongtan area on Chongming Island, which was connected up to the mainland by bridge in 2009 and is now just 45 minutes away from the city centre.

- Urban development, nature conservation and agroproduction (2,700 ha) are being developed simultaneously and on an integrated basis on Dongtan (8,600 ha).
- Dongtan is designed to become the eco-city of Shanghai, with a key focus on sustainable development.
- Greenport Shanghai is one of the elements of the eco-city and is intended as one of the new hubs of metropolitan agriculture (see illustration).

Greenport Shanghai covers an area of 27 km^2. The agropark consists of five elements that are being developed as an interrelated and mutually reinforcing whole:

- Agroproduction park
- Business park (food processing, logistics, trade)
- Central Processing Unit (CPU)
- Research, Training & Education Centre
- Park management

Artist's impression of Agropark Shanghai layout
(From: Masterplan Greenport Shanghai)

Artist's impression Central Processing Unit (Masterplan Greenport Shanghai)

The masterplan outlines four possible scenarios, together with the key figures. The key figures for scenario 2, 'Large-scale', are provided below by way of illustration.

	Sector	Scale	Output
Animal production	Pigs	1 million meat pigs and 120,000 sows	240,000 tonnes of pig meat
	Poultry	2 million laying hens	50 million eggs
		5 million broilers	60,000 tonnes of chicken meat
	Aquaculture	Not quantified	
Vegetable-based production	Vegetables	400 ha of greenhouse	24,000 tonnes of vegetables
	Mushrooms	38 ha of mushroom cultivation	87,000 tonnes of mushrooms
			260,000 tonnes of mushroom compost
	Open-ground farming	Not quantified	

Agroproduction park

This scenario is based on intensive production units operating to Western sustainability standards.

Business park (processing, trade & logistics)

The present shortfall in primary production in the Shanghai region and the problems further down the chain are leading to shortages. These include quality, storage and processing capacity, distribution, the protection of intellectual property, product certification, chain integration, processing and trade. The chain integration, in particular, makes it possible to commercially exploit the high quality of the agricultural products from Greenport Shanghai in the Shanghai market. For this reason 70,000 m^2 have been set aside for agro-processing firms, trade and logistics. The warehousing facilities are a major factor in the huge improvement in agro-logistics in and around Shanghai.

Central Processing Unit

The model for the biofermentation plant in the CPU is based on 50% of the input as described in scenario 2. From this model it follows that 3.4 million tonnes of biomass needs to be available in the agropark. The model is therefore based on 1.7 million tonnes of biomass (approx. 50% chicken and pig manure and 50% other biomass flows).
Such a plant would have a capacity of 50.48 MW and could produce 443 million kWh of electricity, sufficient for 70% of the total energy requirement if the agropark were to be fully utilised. As soon as occupancy levels in the park begin to rise, a second (e.g. in 2014) and if necessary a third plant could be added to the cascade. The surplus electricity could be fed back to the grid. This model does not take any account of these additional plants. In this regard it should be noted that a second plant would be in the interests of the agropark in terms of flexibility and delivery certainty.
The waste-to-energy plant in the CPU is based on a capacity of 460,000 tonnes of waste (either domestic or industrial). The plant has an energetic capacity of 45.7 MW and would produce 400,560,500 kWh of electricity a year.

China suffers from a shortage of qualified staff, while the market is calling for innovativeness and creativity in order to cope with the cut-throat local competition. The concentration of numerous knowledge-intensive companies with Western management creates a dynamic environment in which individual firms learn from and strengthen one another in the development of new products.
On account of the lack of familiarity with the new products in modern technology among potential Chinese customers, demonstration and training are required in order to sell the products. The park also offers space for presentation, training and research.

Park management

As noted previously, the park management is vital for organising 'greater than the sum of the parts', in both a technical sense, via the CPU, and an organisational sense, and for the management of the agropark. Good contacts with the government are highly important for successful business operations in China. In addition the Chinese are used to thinking in big numbers. In the case of large-scale projects, positive publicity and an impressive presence are extremely important for the continuity of the business operations.

The concentration of strengths and sharing of overheads provide added value for the individual agro-businesses. One of those activities is branding. In China this is generally performed by big players. A company such as Yurun Food Group positions its products as hygienic, fresh, nutritious, convenient and modern. Consumers by brand articles as they believe them to be more reliable with regard to such features as food safety and product quality.

The agropark is better placed to position its products broadly as a reliable, locally produced quality product of Western make with food safety certification. Joint promotion by means of TV advertising, shop samples and price strategies can help promote sales more widely.

2.4 The added value of Greenport Shanghai

The sustainability performances

People
Food supply

The rapid urbanisation in China is leading to a growing demand in the big cities for food of consistently reliable quality. The rising living standards have made the rapidly growing middle-classes more particular about agro-products. The park produces quality products that satisfy the rising standards of the middle-class in the city and that are available on a sufficiently consistent basis for the growing urban population.

The production and processing of agro-products on a large scale coupled with high-grade knowledge and modern technology are resulting in efficient methods of production, high productivity and cost benefits.

Consumer confidence

In recent decades Chinese consumers have taken fright over a number of incidents that have undermined confidence in public health, food safety and food reliability (e.g. SARS, melanine, BSE, avian flu, and agricultural pesticides). This has resulted in a fall in local consumption of agro-products. The incidents have also been damaging for international trade, as countries have sealed off their borders by means of veterinary and phytosanitary measures.

The agropark offers a solution in that a transparent tracking and tracing system has been set up with the aid of Dutch knowledge and technology so as to guarantee better food safety. The presence of various large-scale agro-businesses in the vertical and lateral industrial columns in the one central location has made the control and enforcement of the regulations a good deal easier. This provides a guarantee for food safety, certainty of food supply and delivery certainty for partners.

Position of local farmers

Small-scale production in China is organised differently from in the Netherlands. A farming family is assigned a small plot of land by the village community for it to cultivate. This land is used intensively and every three to five years the farmers shift to a different plot, thereby also making for crop rotation. For ease of marketing, all the farmers in the village cultivate the same crop, so that they always have sufficient stocks of the products for dealers seeking to buy.

More and more wholesale companies are discovering China. In order to guarantee the quality and food safety of the products they are looking for direct relationships with producers. With an average farm size of 1,000 m^2, far more producers are needed in order to supply a sufficient volume, particularly if output per m^2 is very low. This has adverse consequences in terms of retailers' stringent product requirements in such areas as uniformity, food safety and logistics. The marked inequality between the rich cities on the east coast and the poor rural areas in the hinterland is a major concern for socio-political stability in China. The central government wants to prevent the mass migration of poor rural dwellers to the big cities by providing that group with a better future. Its aim is to modernise the agricultural and animal husbandry sectors and to restructure the landscape.

With the 11th Five-Year Plan, covering the period 2006 – 2010, the central government is seeking to lay the foundations for the reform of the agricultural sector. The

	Agropark	Biofermentation	Waste2Energy	Total
Total investment	€ 80,990,318	€ 74,884,615	€ 168,500,000	**€ 324,374,933**
Equity	€ 20,247,579	€ 18,721,154	€ 42,125,000	**€ 81,093,733**
Discounted cashflow method	€ 36,743,626	€ 18,928,860	€ 17,448,085	**€ 73,120,571**
Payback period	5 years	10 years	11 years	**9.3 years**
Internal profitability	22.3%	13.3%	11.8%	**14.8%**

aim is to boost modern agriculture, establish a relationship between industry and agriculture and city and countryside and to increase living standards in rural areas. In order to achieve this the government has a number of focal points in mind:

- Accelerating the growth of animal husbandry and assuring the delivery of animal products.
- Improving the supervision of product quality and achieving more consistent product safety.
- Improving the profitability of the sector and hence farmers' incomes.
- Increasing awareness of environmental protection and the eco-structure.

Animal welfare

A great deal of animal suffering takes place in transportation or is due to lack of space and light. Greenport Shanghai is seeking to minimise transportation in the chain and to provide animals with more space and light. Large-scale production also offers opportunities for the improvement of animal welfare. In China that is regarded as exceptionally progressive, as production there is even more rationalised. Or as a Chinese project partner put it, 'Animal welfare? Can we solve our human welfare problems first please!'

Planet
Multiple use of space

There is a lot of pressure in China on the limited availability of agricultural fertile land and sustainable land use. Greenport Shanghai is organised on the basis of spatial clustering, where various links in agro-business are established in one and the same place. This leads to the responsible use of the scarce space in a densely populated area.

Recycling of raw materials

China is grappling with a shortage of animal feed, water and land. The high-efficiency of modern agro-businesses generates savings in raw materials while preserving the high productivity.

China faces a major problem of environmental pollution (surface waters, soil, methane) from bio and other forms of waste. An economic loop in Greenport Shanghai ensures that bio-waste (manure, peelings, foliage and haulm, etc.) are converted in a biofermentation plant into renewables such as electricity, water, heat, CO_2 and by-products. These products then reappear as input for the agroproduction in the park. Any surplus new products (heat, electricity and compost) are delivered to the city. Production is CO_2-neutral.

Sustainable energy

China is facing a growing demand for energy. The price for energy (gas and oil) has risen sharply in recent years (with a temporary dip) and is expected to remain high.

A biofermentation plant in Greenport Shanghai is an important source of cost-savings and is generating a large potential energy surplus. A biofermentation plant will be more profitable at high volumes, meaning that it needs to be on a sufficient scale and to have enough capital.

Water management

China suffers from water shortages and surface waters are subject to pollution from the use of organic manure to fertilise agricultural land and from the overuse of pesticides. Greenport Shanghai provides a solution through the efficient use of water in primary production and the delivery of water from the conversion of biomass into energy and water.

Profit

For entrepreneurs in the agropark

The chain-based system of organisation in Greenport Shanghai brings down the cost of the end-products and improves the utilisation of scarce factors of production due to the efficient use of nutrients, water and energy. The presence of various large-scale agro-businesses in the vertical and lateral industrial columns in the one central location has made the control and enforcement of the regulations a good deal easier. This provides a guarantee of food safety, certainty of food supply and delivery certainty for partners.

The vertical and horizontal clustering of businesses means that an agropark operates at a logistical advantage, with fewer movements within the value chain. Much road transport is eliminated, the transport costs are lower, pressure on urban traffic and environmental pollution is reduced and product quality rises (due for example to less transhipment). The concept also offers the advantages of one-stop shopping for supermarkets.

The concentration of agro-logistics makes for a reliable, consumer-ready fresh product combined with a wide range of products for sale to large-scale customers. Agroparks enable the relatively small SME businesses to establish a sustainable relationship with large retail chains and provides them with the means to be more in charge.

For park management

The spatial clustering of various enterprises in the agrosector provides stable input for the CPU and the stable offtake of the CPU's products. The utilisation of input streams from outside the park and the sale of CPU products within the park increase the return on capital of these facilities, which demand a comparatively high investment.

The quality standards of Greenport Shanghai are also translated into the fee paid by the agricultural entrepreneurs in the park for the services of the park management and in the rental or sale value of the land and/or buildings.

The Renewable Energy Law, which came into force in 2006, provides for the encouragement of alternative sources of energy. The aim of this law is for 10% of energy generation in 2020 to come from renewable sources, such as wind, solar energy and biomass. Among other things the law provides for an energy subsidy of 0.025 euro per kWh on condition that the project is approved by 2010. The law also regulates the 'green' electricity offtake by the government. The Clean Development Mechanism (CDM) programme, which regulates the reduction of greenhouse gas emissions, enables industrialised countries to purchase emission rights (CO_2-credits) from developing countries.

China offers particular opportunities in the field of renewable energy, which Greenport Shanghai taps into.

For knowledge parties

The innovative agroparks concept is to a large extent based on scientific, technical, ecological and spatial knowledge. For the knowledge parties concerned the projects based on this concept are of commercial interest as they can generate large consultancy assignments. In that sense Greenport Shanghai is a successful project as it has generated spin-off in the form of new consultancy assignments, for example in China and India.

For investors

The business plan 'Agropark China' drawn up by Merapeak and Headventure on behalf of TransForum indicates that an agropark can be highly interesting for investors. Separate calculations have been provided in the business plan for the agropark, the biofermentation plant and the waste-to-energy plant. These yield the following:

SWOT analysis of the sustainability performances

Strengths

- Responds to the demand of the rapidly growing middle-class consumers in the city.
- Guarantees food quality and food safety through improved agro-logistics.
- Efficient use of nutrients, water and energy through the organisation of agroproduction.
- Integration of the product chain in a single location, thereby substantially reducing transportation.
- Energy production from waste.

Weaknesses

- High investment costs.
- There is as yet no agropark of this type and size. The technology and synergy in question have not yet been proven in practice.
- A large number of players will lead to slow decision-making and possible disputes.
- In the short term there could be a loss of work for small farms but in the long term the agropark will offer sustainable employment and improved working conditions.

Opportunities

- Value creation from the branding of high quality and safe food.
- If Dutch standards are applied, the 3P performance will improve still further.

Threats/risks

- Small risk, but major impact in the event of animal diseases.
- Linkages and integration can result in greater mutual dependence.

2.5 The value creation model

Notes on the value creation model

Greenport Shanghai stands out on for its intensive production in combination with high sustainability performances. On a relatively small area of land, Greenport Shanghai is producing a large volume of food products. Not just by the spatial clustering of various agro-businesses but also by arranging for integration and linkages, the park is achieving very high 3P efficiency. In addition the Dutch method of production and management ensure far better quality and safety guarantees. It is vitally important for the agroparks concept to be actively monitored and to be actively directed in terms of 3P performance. The park management is therefore indispensable, and enables the overall results of the agropark to be greater than the sum of its parts. The creation of values is shown in the model on the next page.

2.6 From plan to investment

The business plan makes it clear that investing in Greenport Shanghai is an attractive proposition for various groups of investors:

1. Sufficient profitability (private equity/venture capital/banks/institutional investors).
2. Green image (energy companies).
3. Market entry in China, for which a certain scale is required (e.g. waste-to-energy firms).
4. Sustainability and innovation (government support).
5. Infrastructure and sustainable real estate (project developers).
6. Stimulation of innovation, market positioning and cost advantage (individual agro-entrepreneurs).

Even so, the project did not succeed in generating actual investments. To some extent this was due to the economic crisis, which put a damper on large-scale investments in general. But the most important reason concerned the fact

The creation of values may be depicted in the model as follows:

Competencies

- Permanent development
- Chain and network management
- Professional entrepreneurship

*Multi-layered governance (1+1=3)
Training of operators and
professionalisation of entrepreneurs
and developers*

Investments

- Knowledge of and expertise in 3P efficiency (park management)
- Chain management/ logistics/marketing

Project development +
- *Park management*
- *Business development*

- *Concept monitoring (3P) (continuing to invest and further development of business model)*
- *Market development*

Value creation model
Greenport Shanghai

Unique Selling Point (USP)

- Closed, big volume, smallest possible surface area, market-driven

Integration and linkage

Results

- 3P efficiency
- High production
- High quality and safety

that nobody was prepared to assume responsibility for the 'greater than the sum of the parts' aspect. Organisationally, there was no one to undertake the role of 'project developer plus' and to take commercial responsibility for the network management, spatial development, concept monitoring and services required to turn an agropark into a success.

Particularly when the theoretical element of the project was in full swing, so many players, ranging from regional governments to knowledge institutes and from intermediary organisations to agriculture entrepreneurs, wanted to score with the project that the orchestration role was not entrusted to anyone. On a number of occasions the collective interest lost out in the project to individual interests and the drive to score. It is also clear that the way in which the project was approached tended to be dominated by knowledge-players, while entrepreneurs were not sufficiently involved. This meant that there was too much emphasis in the early stages on the spatial and technical design, with the business case being a poor relation.

Finally reference needs to be made to the fact that cultural management is crucial for doing business in China, and this was an area in which very few of the partners concerned had any experience.

2.7 The lessons for the entrepreneur

The entrepreneur as project developer

The plans were not subjected to a sufficient commercial and financial reality check during the masterplan stage of this project. Entrepreneurs and investors had little if any involvement at this stage and, when they were, often as consultants rather than as venture capitalists. Not all the plans proved realistic at the investment phase.

It is also clear that the masterplan cannot be regarded as a 3P business case. The lack of such a case makes it extremely difficult, if not impossible, to generate investments immediately after the masterplan stage, as the Chinese players wanted.

An important lesson is that the development of a 3P business case needs to be spearheaded by entrepreneurs who are putting up venture capital and that knowledge players can play a highly valuable supportive and underpinning role.

The entrepreneur as coach

Although the Greenport Shanghai agropark concept assumes a horizontal, dynamic and development-oriented approach, the Chinese context is to a significant extent still dominated by a hierarchical, top-down and planning-based culture. This made it difficult to set in place the right conditions for the success of the project. Whereas in the Netherlands the use of process monitoring and reflection can help focus on the many 'wicked questions' with regard to power, responsibilities, role perceptions and so on, this turned out to clash with Chinese culture. This represents a major challenge in terms of intercultural management.

The entrepreneur as strategist

The three strategies of metropolitan agriculture come together in Greenport Shanghai. 'Sustainable Intensification' arises as a strategy since characteristic features of the primary production in the project are increased output, increases in scale and more sustainable production methods. At the same time, the strategy of 'Sustainable Valorisation' is in evidence, particularly on account of the marked focus on improved logistics. This logistical innovation is a vital precondition in the Chinese context for the realisation of the high quality standards set for the end-products.

Finally the 'Sustainable Diversification' strategy comes into play on account of the specific link with energy generation, sustainable waste processing and water management. These mean that the agropark has a pivotal role to play at regional level, in terms not just of food production but also of sustainable agro-services.

The linking up of the strategies makes Greenport Shanghai an exceptionally ambitious project. Some of the project participants have argued that a phasing of the strategies would have produced a more realisable plan. Or as one of the project participants put it: 'The masterplan shows how the park will be in 30 years' time if we travel at 300 km an hour. But right now we're standing still, so let's start by talking about first gear.'

The entrepreneur as games-leader

The aspects of Trusting, Explaining and Expecting were not given enough attention in this project. Particularly among the Dutch participants there was insufficient trust and they begrudged each other their successes. Many of the players focused on their own interests and opportunities instead of the collective interest and the need to act jointly. A clear example of this was the competition among the Dutch players for the project leadership. This behaviour was significantly reinforced by the political agendas of the provincial officeholders, who were largely driven by the need to 'score'.

Similarly expectations were poorly managed, especially at the start of the project. During the masterplan stage, for

example, the project leadership voiced its expectations that the Chinese players would be prepared to pay for knowledge, whereas later in the process it turned out that the Chinese were expecting that the Dutch would be investing financially themselves.

The original expectations turned out to be largely assumptions that had not been sufficiently validated by the Chinese partners. Nor was it clear how the decision-making in the project had been arranged, so that a particular stakeholder often either felt unheard or did not feel bound by the agreements reached.

The steering group that was set up failed to do any genuine steering but sought to mediate and to cover up differences. This led to a situation in which the input and goals of the various parties were not sufficiently explained. In these circumstances hidden personal agendas often took precedence over the collective agenda.

In an important lesson is that the three aspects of Trust, Explanation and Expectations can reinforce one another not just positively about also negatively. Clear-cut leadership based on a mandate from the various stakeholders is vital for safeguarding these three aspects.

The entrepreneur as spider in the web

It is abundantly clear that the world lacks a development organisation with the knowledge, expertise, entrepreneurial skills and network to commercialise the agropark concept in the world market. Such a development organisation could create many new opportunities for both agro-entrepreneurs, entrepreneurs from the sustainable energy sector, entrepreneurs from the waste-to-energy sector and research institutes. The scale and complexity of the agroparks concept render a specialist organisation of this kind a necessity. On account of the complexity this will always be a network organisation, for it is inconceivable that a single player would be capable of covering the entire scope of an agropark.

Only through good network management will such an organisation be capable of achieving the necessary synergy.

The entrepreneur as winner

In terms of sustainable, large-scale agroproduction, the agroparks concept at the heart of the Greenport Shanghai project is without doubt highly promising. The fact is, however, that at this stage it remains a concept on paper. While there are potential sustainability performances, there is no successful track record. Nevertheless the economic profitability of the concept and the value creation in terms of people and planet are evident from 'Agropark China' business plan.

The project did, however, lead to spin-off in the form of new consultancy assignments in China and India on the basis of the agropark concept for a number of the chain players concerned. In that sense the project may decidedly be regarded as a success.

2.8 The present challenges

Many of the parties concerned are still left facing the original challenge that confronted them when they got into the project. The Chinese government still faces the task of producing sufficient safe and high quality food for its rapidly growing urban population. For the research organisations it remains important to implement the theory of agroparks in practice.

And for many Dutch agro-entrepreneurs the challenge is to commercialise their leading position in terms of knowledge, expertise and entrepreneurship concerning 'Sustainable Intensification' in a world market.

The project has indicated the pressing need for a new 'developer plus' capable of implementing and monitoring the potential for commercial synergy in agroparks. Who will be prepared to step up to the challenge?

*The entrance of Greenport Shanghai
(Masterplan Greenport Shanghai)*

Beef cattle are able to graze in the Schoonebeekerdiep stream valley of the Koe-Landerij (Photo: Mugmedia, Wageningen)

3. KOE-LANDERIJ

large-scale ánd sustainable

3.1 The challenge

Dairy farmers find themselves facing changes. The increased importance of market forces in the EU Common Agricultural Policy (e.g. cutting back income support and milk quotas) is generating increases in scale in dairy farming. 'Urban' values are becoming dominant in rural areas: the public now assesses agricultural entrepreneurs in terms of animal welfare, sustainability and the landscape. Urban dwellers have a romanticised picture of dairy farming: small-scale family farms with cows in the meadow. An agricultural entrepreneur wanting to scale up must actively earn public acceptance or otherwise face protests and lengthy, expensive licensing procedures.

The brothers *Bouke Durk* and *Berend Jan Wilms* want to scale up and to use the cost savings in order to invest in people and planet. What sets them apart is that they want to scale up in such a way that their farm will contribute towards and blend in with the community in which they work and live.

3.2 How did the innovation come about?

The mission of the Koe-Landerij
'Developing a sustainable, economic perspective for the dairy industry, with public support.'

Inspired entrepreneurs seek link-up with knowledge

The story of the Koe-Landerij ('Dairy Estate') starts with inspired entrepreneurs taking a broad view. Working in a kibbutz in Israel, one of the Wilms brothers saw open-plan dairy sheds at first hand. In addition the Wilms are socially, politically and administratively active in the region. Based on their social involvement, professional interest and inspiration, they took part in the Wageningen UR project 'Cow and Entrepreneur in Balance'. This gave rise to the 'Community Dairy' concept: the basis for the Koe-Landerij that the Wilms brothers are introducing on a regional basis.

They then introduced their idea into the TransForum project 'Dairy Adventure'. Within this project innovative dairy farmers and researchers exchange knowledge concerning their projects: 'Cowmunity' (large-scale farming system), 'Ko-alitie' (family farms cooperation plan in the Province of Friesland), 'Large-scale in small-scale landscape: around Lochem and the Northern Friesian Woods' and 'Cowfortable' (animal-friendly housing system). This knowledge formed the basis of the Koe-Landerij initiative.

The entrepreneurs hold the reins

Even though the project was funded at the planning stage by the Province of Drenthe and TransForum, the entrepreneurs clearly remained the principal throughout the process. In terms of that role they participated in all the project group meetings, conducted the outside meetings and held the presentations. All this reflected the fact that ultimately it is the entrepreneurs who will be investing their time and money in the realisation of the business plan. At the outset the entrepreneurs formulated their underlying principles for the Koe-Landerij:

- Cow at the centre
- People at the centre
- Interaction with the environment
- Public values
- Public return

Project partners
Wilms brothers, Municipality of Emmen, LTO, NAM, Province of Drenthe, Wageningen UR (Livestock Research and LEI), Velt en Vecht Water Board and TransForum.

TransForum project
2008-2010

1880	Wilms Farm established in Schoonebeek
1984	EU introduces milk quota
2005-2007	Wageningen UR project 'Cows and entrepreneurs in balance': Wilms bros. leading participants in the elaboration of Community Dairy
2007	EU announces abolition of milk quota in 2015
2007	Wilms bros. work out Koe-Landerij idea on their own initiative and establish regional contacts
2008	Start of Dairy Adventure project
2009	Dairy Adventure workshops (including Koe-Landerij)
2010	Consortium draws up design and business plan for Koe-Landerij and Voer-Landerij, Province of Drenthe and Municipality of Emmen participate
	Workshop with arable and livestock farmers from the local region
	Workshop on Koe-Landerij and Voer-Landerij strategic communication
	Follow-up workshop with arable and livestock farmers; launch of Agro-, Voer- and Koe-Landerij
	Kitchen-table meetings with local community and presentations to organisations and municipality by entrepreneurs and project team
	Koe-Landerij business plan finished

Governments contribute ideas from the start, on the basis of trust

The Province of Drenthe (rural development) and the Municipality of Emmen (economic affairs) were represented on the project team from the start. This was on the understanding that the civil servants would contribute ideas towards the project, without however formally pinning down the government at that stage. The civil servants in question also took the lead in their own departments, informing their colleagues and political office-bearers. Among other things they informed the project team about the various positions their own organisations were taking and on the required procedures. Confidence and confidentiality within the team were vital, so that those participating could contribute ideas 'freely'.

Emphasis on strategic communication

Together with an expert in strategic communication, all the individuals and organisations that were capable of obstructing or supporting the project were analysed at the outset. A communication plan was then drawn up in order to describe the Koe-Landerij story, with attention to each individual target group.

A great deal of time and energy was then invested in meetings with arable farmers and other dairy farmers in order to discuss new concepts, presentations to be held

Loose housing with compost and turf bedding (Photo: Paul Galama, WUR-Livestock Research, Lelystad)

for governments and organisations and meetings with neighbours and other stakeholders.

Project still at the planning phase

At this point (beginning 2011) talks are still being held with the municipality on the necessary amendment to the zoning plan and all the stakeholders are being kept fully informed.

3.3 Key figures

- The price farmers receive for their milk fell in 2009 by 31% to around 27 eurocents per litre; in August 2010 the price had risen again to 33 eurocents. Milk prices are expected to fluctuate sharply upon the abolition of the milk quota in 2015.
- Average dairy farm size in the Netherlands rose from 42 cows in 1990 to 80 in 2009.
- In 2009 there were 20,000 dairy farms in the Netherlands, a third fewer than nine years before. 4,092 farms had over 100 dairy cows, an increase of 8% (315) on 2008. There are at present 107 farms in the Netherlands with sheds for over 250 cows.
- The biggest dairy farm in the Netherlands has 1,150 dairy cattle and 750 young stock.
- The Koe-Landerij is aiming at 1,000 cows, in a cluster of eight sheds with 125 cows each. Each shed is divided into herds of 50-60 animals. Total annual milk production is 9,000 tonnes.
- The first shed for young stock is to be built in 2012. Thereafter the farm will grow to a size of 1,000 dairy cattle in the space of 5 to 10 years.
- The roughage and even the concentrated feeding stuffs are cultivated in the region itself, by arable farmers supplying the central 'Feed Estate'. Around 500 ha of arable land will be needed for 1,000 cows, including young stock, and 200 to 400 ha for concentrated feedingstuffs.

3.4 The added value of the Koe-Landerij

The advantages of the Koe-Landerij
- A sustainable, profitable dairy farm even after abolition of the milk quota.
- An animal-friendly example of large-scale dairy farming.
- Employment and other services (education and training, children's farm, landscape management) for the region in the agricultural sector.
- New sales channel for arable farmers and potential to expand acreage.
- Increase in scale that does not impair the landscape but in fact supports it.
- Reduction in cost.
- Dairy farming based on regional resources: regional production of roughage and concentrated feedstuff and manure sales.

The sustainability performances

People
Contribution to the community
The entrepreneurs in the Koe-Landerij consider it important for their farm to be of significance for the community. To begin with they do so by contributing towards and maintaining the agricultural sector through the establishment of a profitable dairy farm offering four to eight jobs.
A special feature is that the farmworkers will work no more than 40 hours a week, with the potential for days off. This will make working on a farm more attractive for young employees. The Koe-Landerij will also offer opportunities for training (as a learning business) and opportunities for employees to specialise.

A 'Friends of the Koe-Landerij' organisation is to be set up, in which local residents will be able to contribute ideas concerning the layout and management of the landscape

Koe-Landerij

Increases in scale in livestock farming while retaining the human dimension and meaning for the local community. What does this involve?

➕ The advantages

4-8 employees

➕ Employment

9,000 tonnes of milk

➕ Sustainable, profitable dairy farming

1,000 dairy cows and young stock

4 loose housing

➕ Cost-reduction

125 cows per shed in herds of 50-60

➕ Animal-friendly

➕ Sales channel for arable farmers and extension of area

Compost bedding
The manure is composted in the shed by aeration

500 ha roughage + 200-400 ha concentrated feedstuffs

20 ha

Potential for grazing in the Schoonebeekerdiep stream valley

➕ Supports landscape development

Arable farming
The roughage (and sometimes also concentrated feedstuffs) is grown by arable farmers in the region. Some 500 ha of arable land are needed for a thousand cows. The aim is to reduce costs for both livestock and arable farming.

Arable farming becomes more extensive by renting fields from dairy farms and through greater crop rotation. The input of people and machinery is optimised.

Composted manure for arable farming

Local residents can contribute their ideas for the layout of the landscape park and are able to visit the farm (for recreational purposes) and to buy products.

park to be developed along with the farm. The park will provide space for footpaths and a children's farm, etc. The dairy sheds will be open to the public and local residents will be able to buy milk direct from the farm.

Animal welfare

Operating on the principle of a large-scale farm with 'housed' cows, the aim is to achieve the highest possible animal welfare. On the basis of behavioural research, it has been decided to work with herds of no more than 60 cows. The animals will be held in an open-plan shed with natural compost and turf bedding. The possibilities of free-range facilities are being investigated.

Landscape development and management

The Koe-Landerij will be located in a park-like environment of initially 20 hectares. The landscape development will wherever possible reflect the wishes of the local community. Specialist architects are being engaged to design the dairy sheds so that they blend in with the landscape. Talks are also being held with the Water Board 'Velt en Vecht' about the possibility of allowing beef cattle to graze in the winterbed of the Schoonebeekerdiep.

The establishment of the Voer-Landerij feed centre will enable dairy farmers to contract feed storage out to a regional feed centre. The grounds will then look more attractive.

Planet
Further closing of the minerals loop

The minerals loop, from animal feed to milk, is to be closed further at regional level. The roughage will be regionally produced, while the manure will be composted in the sheds for sale to the arable farmers. On the basis of initial emissions using 'compost bedding', the ammonia emissions in the sheds will be comparable with those in cubicle cowsheds. The dairy industry is currently the biggest customer for soya meal in the Netherlands (300,000 tonnes per year). Once concentrated feedstuffs are produced locally, the Koe-Landerij will no longer have a requirement for imported soya. The Koe-Landerij, the regional feed centre ('the Voer-

Landerij'), the arable farms and any other agricultural activities will together form a 'farm based on regional resources', known as the 'Agro-Landerij' or Agro Estate. Despite extra transport movements, the regional cooperation will bring down total energy consumption.

Greenhouse effect

The net impact on greenhouse gas emissions of the soil and bedding processes in the open-plan shed is not yet clear.

Improvement in soil fertility

If European support for starch potatoes is cut, the arable farmers will need an additional crop. More crop rotation would improve the soil fertility and hence output per hectare. Soil fertility could also be improved if the dairy farmer is able to supply good quality manure with a high organic-matter content.

Contributions to biodiversity

The Koe-Landerij can help foster biodiversity by means of the grazing of the Schoonebeekerdiep winterbed and by including nature development in the layout of the landscape park around the farm.

Profit
Cost reduction and increase in operational certainty

Cost reduction for both livestock and arable farmers is the goal. Dairy farming can concentrate on the further professionalisation of milk production and sustainable animal husbandry.

Arable farming will become more extensive by leasing land from the dairy farm and by more diversified crop rotation, thereby optimising the input of people and machinery. The prospects for arable farmers improve as they become less dependent on growing starch potatoes for the company AVEBE and on support from Brussels.

Local milk production by the farm

Direct milk sales from the farm will increase margins.

SWOT analysis of the sustainability performances

Strengths

- The Koe-Landerij provides a guaranteed outlet for roughage grown by arable farmers.
- The use of the composted manure on the fields improves soil fertility.
- The compost bedding loose housing is better for animal welfare than traditional cowsheds.

Weaknesses

- No weaknesses have been found in the business model.

Opportunities

- Development of biodiversity, the landscape and nature in the 20 ha landscape park.
- Research into ammonia emissions could show lower levels of such emissions on the Koe-Landerij.

Threats/risks

- Investments in sustainability (for example in a composting system for manure) could be at variance with the returns if milk prices were to fall heavily upon the abolition of the milk quota.
- The increase in transport movements could reduce road safety.
- Further research is needed into the emission of greenhouse gases in the animal housing system and into the energy consumption for composting in the cowshed and for transport.

3.5 The value creation model

Notes on the value creation model

The Koe-Landerij differs from a traditional dairy farm given the special emphasis on animal welfare and the environment, the attention to cost reduction and the new type of operations. The aim in doing so is to create more

margin and gain public acceptance. In order to achieve these results the Estate has to cooperate with other dairy farmers, arable farmers and contract workers. Among other things this cooperation translates into a joint investment in the Voer-Landerij feed centre and contracts for fodder delivery and manure sales. The farmer consequently becomes a large-scale manager.

The creation of values may be shown as follows in the model *(see alongside)*.

3.6 From plan to investment

The Koe-Landerij business model is based on a lower cost price from increases in scale on a single location with a cluster of cowsheds. It will need to be demonstrated to investors that the margin remains sufficient, even if the price of milk comes down and if additional investments are made in the sustainability of the system. Fortunately investors are paying more and more attention to people and planet aspects.

The Koe-Landerij can generate earnings from the following sources:

- Sales of milk to dairy processors (with a higher margin thanks to cost reduction) and home sales.
- Landscape management: the entrepreneurs are in discussion with the Water Board concerning the management of the small river Schoonebeekerdiep winterbed.
- The plan is to incorporate energy generation into the business operations by generating bioenergy and solar energy.
- Financing costs are reduced by issuing shares for the local community.
- Income from ancillary activities (children's farm, reception of visitors, etc.).

The most important factor in the business model is the reduction in the costs of labour, machinery and land. A critically important factor in this regard is the 'Voer-Landerij'

Value creation model
Koe-Landerij

Competencies
- Large-scale management
- Market orientation

From farmer to manager:
- *Training of farmers*
- *Co-creation with local community*

Investments
- Technological progress
- ICT

Transparent entrepreneurship

- *Upscaling*
- *Cooperation with other entrepreneurs*

Unique Selling Points (USP)
- Animal welfare and landscape
- Lower milk and meat production costs
- New business management: outsourcing feed production and combination of meat/milk

Management of factor costs

Results
- Higher margin
- Public acceptance
- Green/blue services (grazing, water storage)

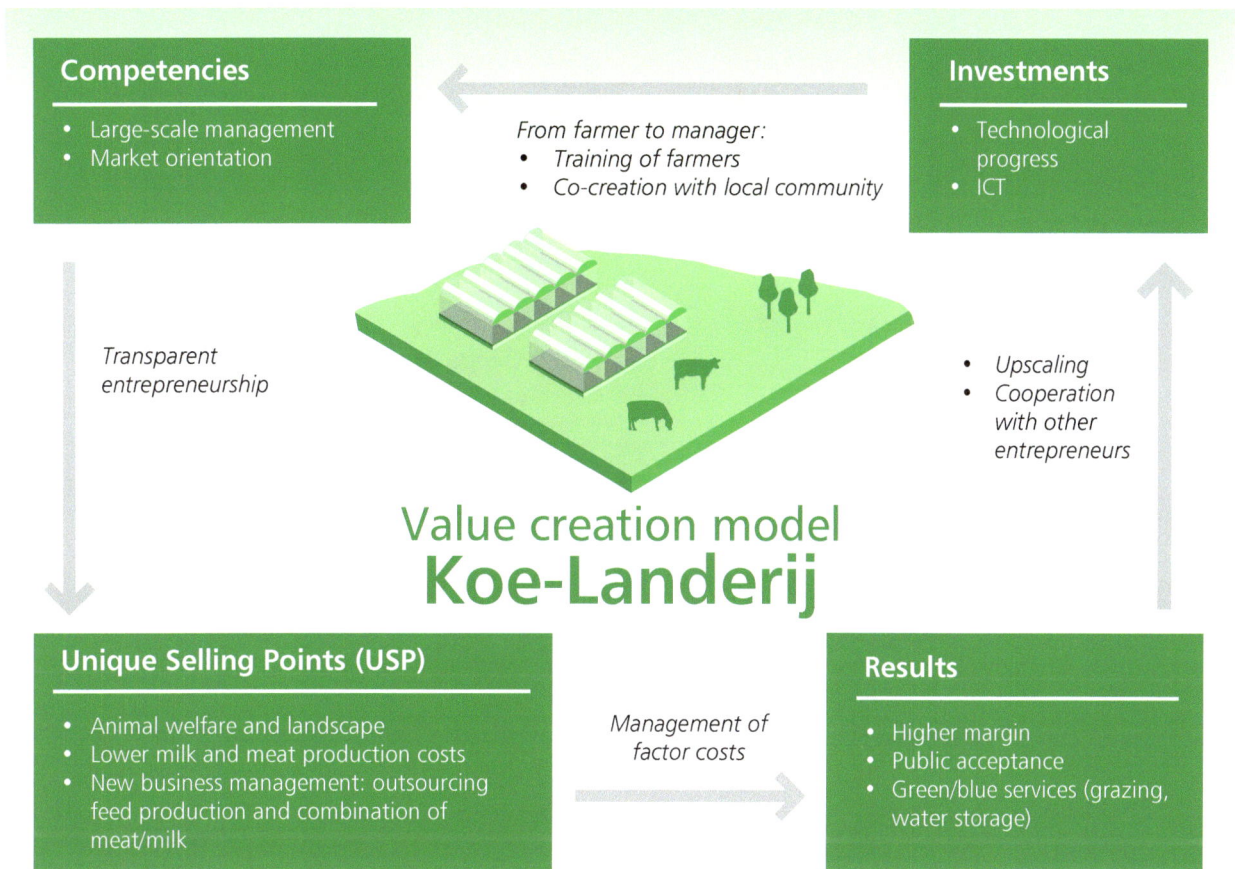

regional feed centre. If the dairy farm can obtain its roughage from arable farmers, it will lower its land costs as it can then rent out or dispose of the land.

The architectural challenge is to design a landscape and farm with optimal logistics to cut down transportation: a design which, with its attractive buildings, also helps enhance the landscape. Public acceptance and efficient operations depend on this. Members of the public also like to see cows in the meadow, as visible evidence of free range.

Risk management

The Koe-Landerij is a growth model, with continuous go/no-go decisions. A start is being made with sheds for 250 cows, growing to 1,000 cows in 5 to 10 years time. By way of fallback option the entrepreneurs already have a right to the construction of a traditional cowshed for 320 cattle. Before the feed centre has been constructed, a start can be made on a direct contract between Koe-Landerij and one or more arable farmers.

3.7 The lessons for the entrepreneur

The entrepreneur as project developer

It was important to involve potential investors and non-governmental organisations such as the Netherlands Society for Nature and Environment right from the planning stage, which is where the project is now. The development of a 3P business plan is currently being worked on.

As this kind of plan is new for those concerned, a workshop bringing together NGOs and potential investors to discuss what they understood by a 3P business plan provided important input. This is expected to simplify the step from 'planning phase' to 'investment phase'.

The entrepreneur as coach

Obtaining experimental status from the government is important for the success of the Koe-Landerij project, as an entirely new housing system is being developed: a 'village' of loose housing for 125 cows, in herds of approximately 60 cows with compost bedding and possibly also free range facilities. This type of housing is more sustainable than conventional large-scale stalls. Since the concept is new, no proven technologies can be used as normally asked for in the case of environmental permits. The first such housing will be experimental and will be the subject of ongoing monitoring.

The entrepreneur as strategist

An entrepreneur who opts for Koe-Landerij is opting for the strategy of *Sustainable Intensification*. A choice in favour of production for the commodity market involves a limitation on the available resources for investment in sustainable development. The Koe-Landerij pursues cost reduction by means of increases in scale and specialisation in animal breeding and care and by manure processing in the shed and management of the feed.

As against this is an anticipated fall in milk earnings after abolition of the milk quota. In order to cover the investment costs there must be a sufficient margin on dairy production for a number of years. If that fails, consideration will need to be given to the production of specialties with a higher value.

The entrepreneur as games-leader

The Agro-Landerij feed centre will only work if livestock and arable farmers act in harness, taking account of each other's operating systems and granting one another advantages. The Koe-Landerij dairy farmers cooperate with arable farmers in the feed centre, and so are dependent on each other.

The entrepreneur as spider in the web

In order to be part of the community and gain public acceptance, communication is a priority. In setting up the Koe-Landerij the Wilms brothers have opted specifically for a participation model: they want to make it possible for people to take part. The Koe-Landerij therefore started with a number of workshops with a strategic communication advisor in which the entrepreneurs drew up a communication plan. Partners in the communication included the local community and societal organisations.

The lesson drawn by the communication advisor was: *'Consult the local community about your plans in good time and take objections seriously. Don't consult once the plans have been finalised but involve the local community right from the start. Do something with ideas and suggestions. Be honest and clear about where public participation is still possible and where it is not. Negotiate as the development unfolds and not in retrospect; otherwise you will generate opposition. By definition people don't like change. People's resistance to large-scale dairy farming is always emotionally based. Always take those emotions seriously.'*

Civil servants from the municipality and the province participated in the project group from the outset. This meant that they contributed ideas towards the practical potential of the project and the relationship between the Koe-Landerij and the policies of the various levels of government. To make it possible for civil servants to participate in this way it is important for the project team to come across to them as 'safe' and 'reliable', and that they are not held accountable for the positions ultimately taken by the municipal or provincial executive or by colleagues.

The entrepreneurs involved researchers from Wageningen University in the idea by taking part in a study group entitled

'Cow and Entrepreneurs in Balance'. The researchers helped plan the new farming concept, the technique and the organisation. In addition experts' knowledge with regard to communication and elaborating business plans was tapped on an ad hoc basis.

The entrepreneur as winner

For the innovation to get off the ground in practice the focus of the project was right from the beginning the development of a 3P business model. The development process consisted of three tracks:

1. Calculating the business model and the sustainability performances.
2. Developing knowledge concerning the new housing system and the business concept.
3. Designing the dairy sheds and harmonising the Koe-Landerij with the local area.

3.8 The present challenges

The Koe-Landerij is still at the planning phase. The most important challenges at the time of writing were:

- Acceptance of the business plan by investors.
- Obtaining the necessary licences and modifying the existing zoning plan by demonstrating that all the statutory conditions (and indeed more) had been met.
- Gaining public acceptance by indicating how the project will harmonise with values of importance in society.

The calculations will indicate whether the intended cost reduction will be sufficient to fund the investments in this sustainable innovation. If that is not the case the entrepreneurs will need to consider a switch from commodities to specialties by adding value to their products. The emphasis on animal welfare and the landscape can be an important supportive factor.

Bouke Durk and Berend Jan Wilms at the proposed site near Schoonebeek (Photo: Mugmedia, Wageningen)

Artist's impression: New Mixed Farm is being combined with innovative architecture (T R Z I N bv, illustration Erik Visser)

4. NEW MIXED FARM
sustainability by closing the loop

4.1 The challenge

New Mixed Farm is a pilot project for an agroproduction park in which loops can be linked up and closed through the combination of businesses from various sectors. Substantial sustainability gains are made possible as a result.

The entrepreneurs in the New Mixed Farm were spurred on by two challenges. The first of these relates primarily to the Netherlands, while the second is to do with the growing urbanisation of the world.

To start with the first of these challenges, the development of agriculture in the Netherlands has given rise in recent decades to intensive forms of animal and vegetable-based production. This intensive production has consequences that have led and continue to lead to major public debate. The issue is therefore one of converting the disadvantages of intensive agriculture into a positive contribution. The entrepreneurs in New Mixed Farm are well aware of this.

The entrepreneurs want to gain public acceptance, where they have the following ambitions:

- Closed loops wherever possible
- More efficient use of (especially regional) raw materials
- Lower emissions of odour, dust and minerals
- No use of fossil fuels
- Reductions in transport by shortening the chain
- Improved animal welfare and animal health
- Creating innovations for agriculture
- Cooperation with the wider community and with each other.

The second challenge has its origin in the fact that more and more people throughout the world are living in cities. The difference between the big cities and rural areas is therefore becoming increasingly blurred. Agriculture increasingly produces for big cities and metropolises; hence the designation 'metropolitan agriculture'. Reliable food chains that deliver sustainably produced fresh products are strategically important in this regard. Agroclusters or agroparks near to the city responds to this concern. Entrepreneurs have been only too aware in recent years that the realisation of an agroproduction park with intensive livestock farming in the Netherlands is easier said than done. This chapter describes their experiences.

An *agropark* is a system innovation
for agroproduction, processing and logistics, based around the clustering of agricultural and non-agricultural functions in various sectors.

An agropark holds out the prospect of closing loops, reducing transport, and efficient land-use.

4.2 How did the innovation come about ?

Back in 2001 the three entrepreneurs – *Marcel Kuijpers* (poultry farmer), *Martin Houben* (pig farmer) and *Peter Christiaens* (contractor) – teamed up in order to build an agropark in the north of the Province of Limburg.

Project partners

Christiaens Engineering and Development B.V., Municipality of Horst aan de Maas, Heideveld Holding B.V., Houbesteyn Holding B.V., Kuijpers Onroerend Goed B.V., KnowHouse B.V., Ministry of Agriculture, Nature and Food Quality (LNV) , Province of Limburg, TransForum, VU University Amsterdam (Athena Institute) and Wageningen UR (F&BR, Livestock Research and PPO).

TransForum project
2004-2009

To begin with the agropark had highly ambitious plans for closing loops. The initial plans paid little attention to the feasibility of the ambitions, especially the commercial implications. Once the entrepreneurs began to put the concept into practice it soon turned out that the dependency on linked businesses entailed excessive risks. For this reason, the ultimate design provided for a jointly managed bio-energy power plant in combination with separate farms.

The original plan, in which the poultry and pig farms, as well as a mushroom farm, would cooperate in the Californië greenhouse area and organise the management of residual and by-products via a joint substances plant came to grief on lack of support from the new Californië greenhouse area. Greenhouse growers were concerned about odour and dust nuisance and had difficulty coming to terms with the negative image of intensive livestock farming. A scaled-down version was then pursued.

In 2006 the participating mushroom-grower went into receivership. The remaining entrepreneurs decided to continue with a robust, open structure, in which other players, on both the supplier and customer side, could link up with the core. From that point on the core consisted of a poultry farm with slaughterhouse, a pig farm and a bio-energy power plant/manure-processing plant. A site was found in a so-called Agriculture Development Area.

Upon commencement of the project, the technology of the New Mixed Farm was expected to be the most complicated part of the project. As matters progressed this turned out not to be the case. The most unmanageable element proved not to be the technology but the social embedding of New Mixed Farm. During the term of the project the social climate in the Netherlands changed considerably. The emergence of the political Party for the Animals is one example of the change in public attitudes towards intensive livestock farming.

One result was that in response to the change in the political configuration, the initially positive stance on the part of the Ministry of Agriculture, Nature and Food Quality – in the form of promised separate status for the project – switched to a more critical one. Similarly the local alderman was required to moderate his initially positive attitude in response to the growth in local resistance. The local elections on 19 November 2009 were also a factor.

Specifically in the case of New Mixed Farm, this was on an unprecedented scale for the Netherlands and involved the link-up of farms and other businesses, while the technology was not yet a proven one. This stacking of innovations meant that the entrepreneurs found themselves facing extra-legal requirements.

Some examples of where this led in the regulatory sphere:
- The strength of New Mixed Farm is that the farms are linked up. In terms of the ultimate licence, however, the stench circles of the separate farms are added together. Two separate licences can be obtained without problems but if they are linked up the requirement switches from no dwellings within a 1,000 metre radius to no dwellings in a 2,000 metre radius from the New Mixed Farm. In this way a sustainable innovation can be obstructed or delayed by the regulations.
- The province indicated that it wanted to operate on the basis of three separate licensing procedures for the three elements of New Mixed Farm, but the municipality wanted the entire procedure to be handled as a single, combined application.
- The Ministry of Agriculture, Nature and Food Quality asked that proof be provided that the technologies in question were genuinely innovative, while on the other hand the Ministry of Housing, Spatial Planning and the Environment said it would withhold approval of the poultry housing system as no practical experience had as yet been gained with it.
- An underground pipeline was planned in order to convey the manure from the pig farm to the bio-energy power plant. This would greatly reduce transport movements and odour nuisance for the local community. A no-brainer, surely? Except that the planned transportation of manure by pipeline was not in accordance with the current regulations, as it did not meet the tracking and tracing requirements based on transportation by road.

The examples also illustrate how difficult it is for SMEs to innovate and how they have to be prepared for the long haul. The lengthy licensing procedure and the supplementary requirements are a direct result of the quantum leap in scale proposed by New Mixed Farm and the growing public debate about the sustainabilisation of intensive livestock farming.

The Environmental Impact Assessment (EIA) was completed in July/August 2010 and the applications for the environmental permit and building permit phase 1 were submitted. At the same time supplementary requirements were laid down. In mid-2010 three supplementary investigations into the health effects, the financial feasibility of the New Mixed Farm and the management skills of the entrepreneurs were still underway. These are investigations that are not normally required.

Communication with the local community

The entrepreneurs devoted considerable attention to communication with the local community. Right at the beginning this was particularly awkward since the entrepreneurs' own plans were still unclear. Whereas the emphasis in the beginning was mainly on the provision of information, at the end of 2008 interaction was sought and working sessions with local residents were organised.

In mid-2008 the debate was – partly through the agency of the political Socialist Party – extended nationally to agroparks in general. That was and is a debate that is too big for the entrepreneurs in New Mixed Farm alone. They therefore decided to provide less of a public face in the second half of 2008. In 2009 a great deal of time was devoted to the preparation of the environmental impact assessment.

A summary of the meetings held up to the end of 2008 is provided below.

8 November 2006	Public information meeting concerning EIA initial memorandum.
10 March 2007	Excursion by the Municpal Executive and Horst a/d Maas Municipal Executive to meet the entrepreneurs and businesses in New Mixed Farm.
26 May 2007	Excursion by Socialist Party representatives of Horst a/d Maas Municipal Council to meet the entrepreneurs and businesses in New Mixed Farm.
4 October 2007	Information meeting for local residents, village committee and political parties.
8 November 2007	Discussion meeting for local residents concerning harmonisation with the landscape.
21 November 2007	Discussion meeting for local residents concerning traffic and transport.
27 June 2008	Visit by local residents to inspect air-scrubber.
26 September 2008	Reception of Limburg Provincial Council committee in response to a citizens' initiative.
6 November 2008	Meeting for local residents concerning harmonisation with the landscape.
19 December 2008	Visit by neighbouring Grubbenvorst Village Council to inspect air-scrubber.

New Mixed Farm does not as yet have any direct, concrete form of cooperation with societal organisations. Locally the entrepreneurs find themselves dealing with the 'Preserve the Pearl' initiative to block the New Mixed Farm. If the differences of opinion are so clear, it becomes difficult to cooperate. There have, however, been a number of evenings on which the entrepreneurs and those behind the 'Preserve the Pearl' initiative have held discussions.

At national level cooperation with, for example, the Animal Protection Foundation could be logical since the poultry element of New Mixed Farm, in particular, offers a number of advantages in terms of animal welfare. Any such involvement will depend on whether this would be sufficient and whether the Animal Protection Foundation would be able to project itself as an organisation that had achieved a substantial improvement in animal welfare in collaboration with an entrepreneur.

Cooperation among the entrepreneurs

It took 18 months for the consortium of entrepreneurs to achieve stability. In 2006 joint working sessions were held in order to work on the New Mixed Farm business plan. This logically lead to a reality check on the ambitious linkages that had been provided for in the planning phase to close the loops wherever possible.

This was logical in the sense that an entrepreneur will wish to avoid unjustified risks and to limit dependencies. What is sustainable from the viewpoint of environmental gains need not by definition be sustainable from the viewpoint of business management, as this example shows.

The joint business planning by the entrepreneurs has not yet been completed. The mutual settlement prices for the products to be exchanged between the various elements of New Mixed Farm still need, for example, to be determined.

Cooperation with the government

Right from the outset of the project it was recognised that obtaining the necessary licences and permits would be an important success factor for the success of New Mixed Farm. The innovative nature and linkage of streams meant that the project represented a licensing challenge.

A task force was therefore set up consisting initially of 13 people, with representatives from the municipality, the province and the Ministry of Agriculture, Nature and Food Quality. In fact New Mixed Farm was asking the various levels of government not to make current or retrospective assessments but to help with the design.

To begin with the separate status promised by the Minister of Agriculture, Cees Veerman, worked well and civil servants were eager to participate and contribute ideas in the task force. That switched fairly rapidly into a defensive attitude. One explanation may have been that at a certain point there was no longer a chairman enjoying political respect, as a result of which the political support for New Mixed Farm ebbed away and it became more difficult for civil servants to support the initiative. In addition the task force was too large to function effectively and the questions too unfocused, so that some of the participants lost interest.

As well as this, the 'separate status' largely became the very opposite of what Minister Veerman had intended at the time. On account of the innovative nature of New Mixed Farm, the farm-size and the public debates, all sorts of extra tests and opportunistic rules were devised with which the project had to comply.

In the case of New Mixed Farm it turned out that the Agricultural Development Area policy lacked significance as it had not yet been operationalised. Plan development needs to be linked to political support and investment. A far-reaching innovative concept such as New Mixed Farm is virtually unable to get by without formally endorsed experimental status in the legislation.

For entrepreneurs wishing to launch comparable initiatives it may be worth concentrating on the following elements vis-a-vis the government:

- Ask the government (in the form of cooperation between local, provincial and national government) to undertake the detailing of the zoning plan and the preparation of the area as also done in the case of other agricultural development areas.
- Ask for cost/benefit and sustainability analyses for a number of different locations so that underpinned trade-offs can be made at local, provincial and national level.

Cooperation between entrepreneurs and researchers

Following initial scepticism on the part of the entrepreneurs, the cooperation between the entrepreneurs and researchers worked particularly well. The researchers made valuable contributions towards the development of New Mixed Farm.

The relevant success factors were:

- Pay sufficient attention to the business aspects of new concepts such as the risks, the earnings models and the financial consequences of various scenarios.
- Regular feedback of research results means that entrepreneurs are able to respond and adjust and have a sense of genuine involvement.
- Entrepreneurial issues generally call for integral responses. Integration among research disciplines is achieved by means of a linking research coordinator and working sessions in which the researchers link up their respective results.

2004	Entrepreneurs join forces in order to establish an agropark in North Limburg to be known as New Mixed Farm
	Agriculture Minister Veerman promises New Mixed Farm separate status
2005	Greenhouse growers drop out as they want to prevent horticulture from being tarnished by the negative image of intensive livestock farming. Decision by remaining entrepreneurs not to admit any new entrepreneurs

for the time being and to draw up a robust business plan Task force appointed with civil servants from various departments to prepare New Mixed Farm for licensing status. Anticipating how this could be facilitated instead of conducting a retrospective test was an innovative step Switch of location by Municipality of Horst from site near Californië to one in the Witveldweg Agriculture Development Area (LOG) Communication with local residents concerning plans

2006	Mushroom grower goes bankrupt and drops out. Link between animal and vegetable-based production is therefore (at least temporarily) lost
	Cooperation among entrepreneurs becomes closer as a result of various working sessions, a trip to China and working on a concrete design and business plan
	Architect appointed to harmonise the commercial activities with the landscape EIA information evening for local community. Local residents turn out to be particularly concerned about the size of New Mixed Farm and its impact on the local environment
2007	New Mixed Farm works at communication and the provision of information by means of farm visits, newsletters and workshops for local residents Emergence of national debate concerning large-scale livestock farming. New Mixed Farm linked up with this, partly by Socialist Party GPs in the local community intensify debate about MRSA and particulates Postponement of decision on LOG Witveldweg by municipality in response to growing opposition

2008	Positive decision by municipal council concerning New Mixed Farm in LOG Witveldweg, narrow majority (+1) Entrepreneurs deliberately opt for news blackout. The disadvantage of this is that support among stakeholders also declines. Independent sustainability scan carried out on behalf of Municipality of Horst demonstrates that New Mixed Farm is a sustainable initiative but has aroused local opposition as poultry farming is new to the area. Friends of the Earth Netherlands 'No to Large-Scale Livestock Farming' citizens initiative rejected
2009	EIA procedure commenced, but is protracted and complicated. Municipality demands supplementary analysis of health effects, financial feasibility and management skills of the entrepreneurs. Positive findings in these areas are a precondition for the sale of the land to the entrepreneurs to establish New Mixed Farm
2010	Submission of the New Mixed Farm Environmental Impact Assessment in July

4.3 Key figures

New Mixed Farm consists of three elements:

- A closed pig farm, a collaborative venture between Heideveld and the Houbesteyn Group.
- Kuijpers Kip, a fully closed poultry firm with slaughterhouse.
- The bio-energy plant for processing the biomass stemming from the pig and poultry farms, etc. This bio-energy plant is owned by the farmers and the Christiaens Group, which has responsibility for the technological side of the plant.

Part 1 A closed pig farm

The pig farm consists of the extension of an existing unit housing meat pigs into a closed pig farm with space for:

- 2,436 empty and pregnant sows
- 45 breeding boars
- 600 sows in lactation
- 10,836 weaned piglets
- 720 breading guilts
- 20,580 finishing pigs

A fodder plant is being built to put together the feed for the animals. The feed mixtures are composed of various components, including wet and dry by-products (base materials, premixes and minerals) obtained from the food industry. These products will be delivered to the farm and processed.

Part 2 A fully closed poultry farm with slaughterhouse

The poultry farm has space for 1,059,840 broilers and 74,448 parent animals from a meat breed. In addition a slaughterhouse is to be built at a farm. A fodder plant is being built to put together the feed for the animals.
The chain for the poultry farm will be fully closed: literally from egg to chicken fillet at the one operating site.

Part 3 Bio Energy Plant

At the core of the physical collaboration between the farms in New Mixed Farm is the Bio Energy Plant. Large streams of animal manure, organic residual substances (breeding waste, carcasses, slaughter waste) and slaughter/waste water will be produced within New Mixed Farm. These streams will be fermented in the bio-energy plant for use as heat and energy and will be processed into compost. The energy will be generated by an on-site Combined Heat and Power (CHP) plant. The fermented manure will then be composted and sold.

The industrial ecology has been established by the various enterprises as shown in the diagram on the next page, whereby the pig and poultry farms have been linked up via manure processing and composting. Divergent scenarios, which could be developed over the course of time, have been worked out for the entrepreneurs.

The basic scenario provides for a combination of co-fermentation and then composting. In the case of co-fermentation the pig manure is mixed with vegetable material (available in the immediate vicinity of Horst). The co-fermentation releases a substantial amount of biogas which, with the aid of cogeneration, can be converted into electricity, with the release of heat and CO_2.
The CO_2 can potentially be used for the 'fertilisation' in the adjoining greenhouse area (normally greenhouse growers use natural gas in order to make CO_2) and the heat can also be used in the greenhouse.

Heat could possibly also be made available for other businesses and/or local residents. The bio-energy plant will ferment between 60,000 and 120,000 tonnes of organic material a year. The precise amount will depend on whether local livestock farmers decide to take part. The remaining digestate will for the time being be consigned to a composting plant.

This will result in a high-grade and clean (mushroom) compost that can also be sold outside the agricultural industry itself and be exported. Partly with the aid of some of the heat from the co-fermentation, the water in the manure is evaporated. It is possible that water could be sufficiently clean for use in, for example, greenhouses. This part of the bio-energy plant will have a maximum capacity of 50,000 tonnes.

The ambition is to take the next step as quickly as possible and to complement the fermentation and composting with incineration or gasification. This extra step will once again release sustainable energy, leaving behind just a small amount of ash that can also be readily sold outside the agricultural industry. The only point is that the techniques in question still lack the required operating reliability.

Share of New Mixed Farm in Dutch livestock population

In 2009 approximately 12 million pigs, 3.6 million cows and 93 million chickens were kept in the Netherlands. Despite the fact that New Mixed Farm brings together a large number of pigs and chickens in the one location, the farm accounts for just a small proportion of the total Dutch livestock population, namely 0.29% of the total number of pigs and 1.29% of the total number of chickens.

New Mixed Farm

Sustainability gains by the closing and linking up of various loops.

➕ The advantages

Poultry farm with slaughterhouse

The loop is fully closed; literally from egg to chicken fillet in the one business location.

broilers	1,059,840
slaughter breed parent animals	74,448

Bio-energy plant

The plant produces compost, fertiliser, heat, CO_2, electricity, biogas and clean water.

Closed pig farm

A loop from the breeding of pigs to meat production in the one location.

finishing pigs	20,580
weaned piglets	10,836
pregnant sows	2,272
other	1,549

➕ **7%** reduction in poultry-keeping costs

➕ Lower animal transport means less stress and improved animal welfare

➕ **60 - 80%** lower consumption of fossil fuels

➕ **30 - 40%** lower emissions of greenhouse gases

➕ **60 - 85%** of the required energy is utilised in the form of renewable energy

➕ Lower animal transport means less stress and improved animal welfare

➕ **60 - 80%** lower consumption of fossil fuels

➕ **30 - 40%** lower emissions of greenhouse gases

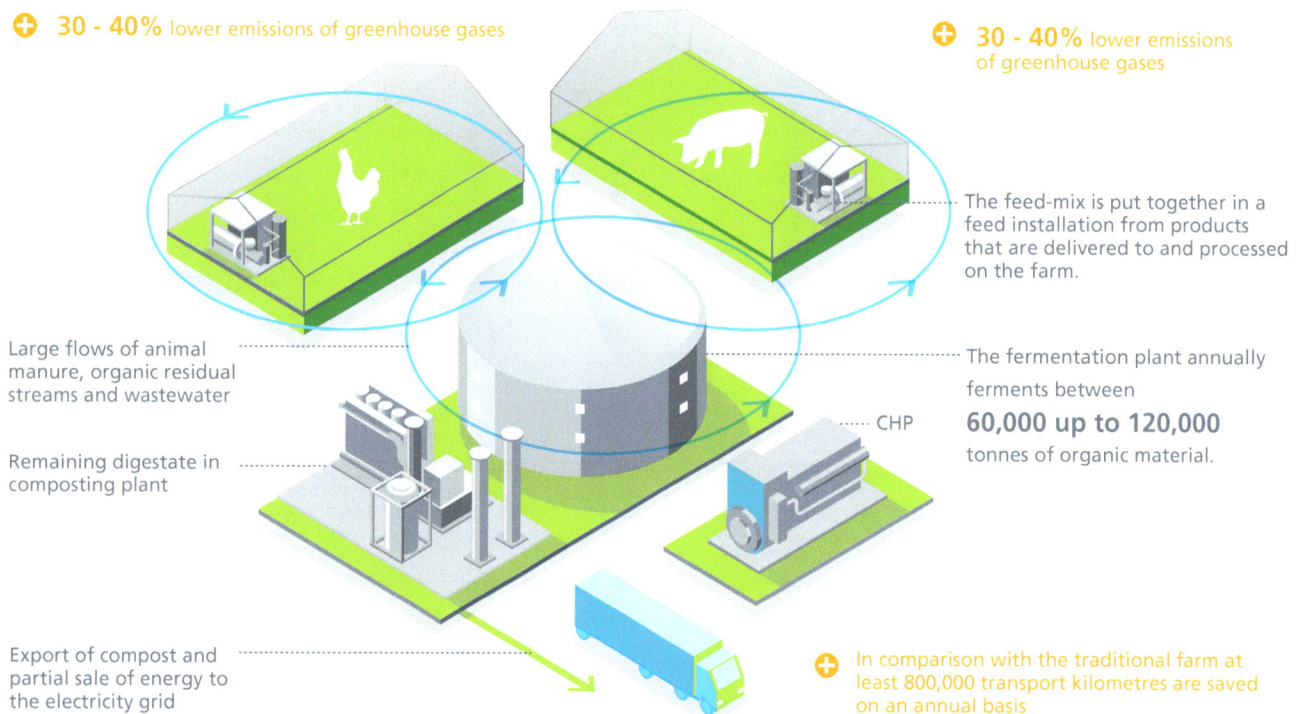

The feed-mix is put together in a feed installation from products that are delivered to and processed on the farm.

Large flows of animal manure, organic residual streams and wastewater

Remaining digestate in composting plant

The fermentation plant annually ferments between **60,000 up to 120,000** tonnes of organic material.

CHP

Export of compost and partial sale of energy to the electricity grid

➕ In comparison with the traditional farm at least 800,000 transport kilometres are saved on an annual basis

4.4 The added value of New Mixed Farm

A brief summary of the most important results of the New Mixed Farm is provided below, divided into people, planet and profit aspects and relevant underlying sustainability aspects.

The actual and potential advantages of New Mixed Farm

- Thanks to the cooperation between different businesses, New Mixed Farm performs well above average in terms of environmental impact. Thus 60-80% less energy is used, emissions of greenhouse gases are down by 30-40% on those in conventional farms and the use of air-scrubbers means that New Mixed Farm, with a 70% reduction in ammonia, has already nearly achieved the policy target of a 75-85% reduction by 2030.
- Thanks to the chain integration on both the poultry and the pig farm there has been a substantial reduction in animal transportation, which has in turn helped improve animal welfare.
- The profitability of the farms is rising, apart from which New Mixed Farm is also providing a boost for the local economy. In the new situation there are some 65 employees.

The sustainability performances

People
Working conditions

New Mixed Farm no longer involves heavy manual labour, such as catching broilers, hanging live broilers on slaughter-hooks and cleaning the hutches (the scale makes it possible to use a cleaning robot). On account of the size of the farms the work will become more specialised. This could be perceived both negatively (less variation) and positively (greater involvement).

Animal welfare and animal health

The poultry element of New Mixed Farm scores better on animal welfare than conventional poultry farming. The integration of production with the slaughterhouse has greatly reduced animal transport and hence also stress. Apart from that, the animals are also stunned before slaughter. In the case of broiler breeders the veranda colony system was used and for broilers the patio housing system developed by Kuijpers Kip in collaboration with Vencomatic. The rearing mortality rate in the patio system is around 1% lower than normal and the output percentage 1% higher.

With regard to an important aspect of animal welfare in broiler housing, namely the use of fast-growing strains, New Mixed Farm does not differ significantly from conventional broiler farming.

The pig farming within New Mixed Farm scores no better or worse than conventional pig farming in respect of animal welfare and animal health.

Health effects

MRSA is encountered in pig and calf husbandry. Infection takes place via direct contact between people and animals. The risk of MRSA infection for local residents is negligible since air-borne infection is highly unlikely. For more information on fine substances see the section on 'Emissions'.

Dissemination of animal diseases

Direct or indirect contacts with other poultry farms are greatly reduced in New Mixed Farm by the integration of the various stages of broiler production in the one location. Needless to say this greatly reduces the risk of infection by animal diseases. The pig farm is a closed unit. The sows that are kept as meat pigs come from sows and breeding boars kept on the same farm. This means that few if any animals are brought in from outside – something that is highly unusual for a farm on this scale in the Netherlands. Various other measures have also been taken on the pig farm to avoid contact with the outside world or between animals at the various stages of production.

These include:

- Application of the clean road/dirty road principle.
- Dead pigs are kept in refrigerated containers.
- Transportation to the slaughterhouse does not involve collecting pigs from other farms.
- Animals in different age-groups are kept separately.

In this regard the road between the various pigsties is used as a natural barrier.

Planet

Greenhouse effect and exhaustion of fossil fuels

New Mixed Farm scores positively in comparison with conventional farming when it comes to reducing the greenhouse effect and the exhaustion of fossil fuels. This is attributable to two elements:

1. The bio-energy plant produces renewable energy by the co-fermentation of manure and co-products. 60 to 85% of the energy required in the chain for feed, transport and animal production is offset by the generation of this renewable form of energy.
2. The reduction in methane emissions by co-fermentation.

Emissions

At local level New Mixed Farm scores neutrally overall. According to the Environmental Impact Assessment, emissions of ammonia, particulate matter and odour are falling at the planned site for the pig farm and are rising at the future location of the poultry farm and the bio-energy Plant.

At national level there has been a substantial reduction in emissions of ammonia, odour and fine substances since the best available techniques (air-scrubbers) are being used to limit emissions via ventilation air.

Manure

New Mixed Farm does not in any way increase the problem of the over-fertilisation of agricultural land. All the manure from the animals is converted into energy or organic fertiliser for export.

Profit

Investments

The bigger size of the New Mixed Farm businesses generates economies of scale. This cuts the cost of meat production (7% in the case of poultry) and facilitates the necessary investments to comply with current and future legislation.

The reduction in cost is due in particular to the lower transport costs and the elimination of handling costs (i.e. the fact that chickens no longer have to be attached manually to the hook in the slaughterhouse). On account of the closed loop at least 800,000 transport kilometres are saved on an annual basis in comparison with a traditional farm.

SWOT analysis of the sustainability performances

Strengths

- Positive effects with respect to working conditions, animal welfare and health and the risks of animal diseases. All this thanks to the integrated poultry production chain.
- Reduction in the greenhouse effect and the exhaustion of fossil fuels on account of the manure-processing and generation of green energy by the bio-energy plant.
- The planet indicators for the environment at national level since the animal production that is now spread all over the country is concentrated in the one location in New Mixed Farm.

Weakness

- There are no sustainability indicators to show that New Mixed Farm outperforms the benchmark.

Opportunities

- Develop criteria for the chain, especially for mixed-feed producers, such as the use of sustainably produced base materials (for example the EKO organic hallmark/Milieukeur environmental hallmark/sustainable soya).
- Improvement in animal welfare of the pigs.

Threats/risks

- The most important critical success factors are the economic feasibility of the bio-energy plant and the chicken slaughterhouse. The economic feasibility of the bio-energy plant depends on external factors such as subsidies and price movements for raw materials for the production of bio-energy and on the market for manure. If the bio-energy plant and/or the chicken slaughterhouse are no longer able to operate, the gains made on various sustainability indicators will also lapse.
- Diminishing public support for large-scale animal production.

4.5 The value creation model

Notes on the value creation model

New Mixed Farm delivers various products: live pigs, chicken meat, organic manure and energy. The first two products are obviously supplied by the pig farm and the poultry farm and the latter two by the bio-energy plant. The latter adds value to manure by processing it into energy and organic fertiliser. Waste is consequently converted into income and does not form a cost item. The energy is partly used on site and partly sold to the electricity grid. The organic fertiliser is exported. In relation to the bio-energy plant it is important to note that the generation of renewable energy from co-fermentation

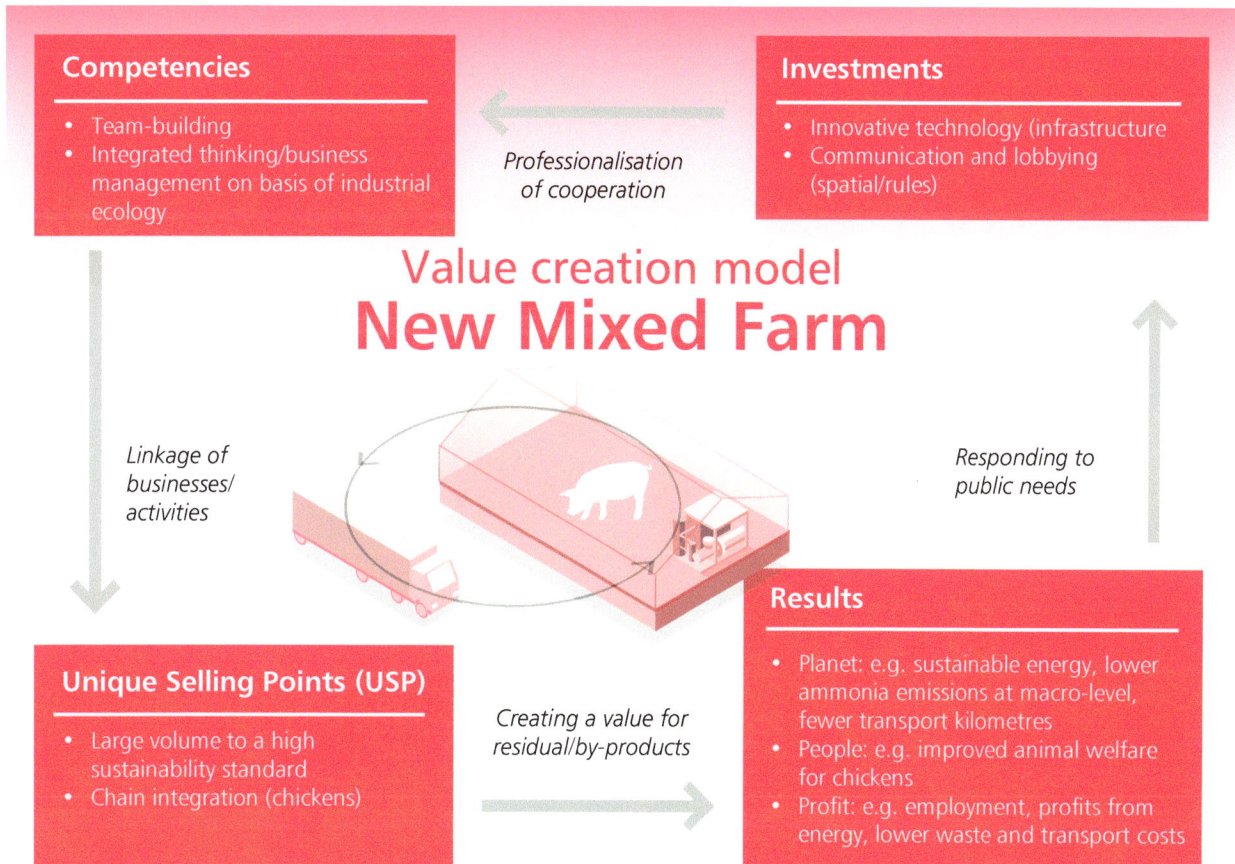

Competencies
- Team-building
- Integrated thinking/business management on basis of industrial ecology

Professionalisation of cooperation

Investments
- Innovative technology (infrastructure
- Communication and lobbying (spatial/rules)

Value creation model
New Mixed Farm

Linkage of businesses/ activities

Responding to public needs

Unique Selling Points (USP)
- Large volume to a high sustainability standard
- Chain integration (chickens)

Creating a value for residual/by-products

Results
- Planet: e.g. sustainable energy, lower ammonia emissions at macro-level, fewer transport kilometres
- People: e.g. improved animal welfare for chickens
- Profit: e.g. employment, profits from energy, lower waste and transport costs

depends on subsidies for its profitability. Based on 2008 market prices the subsidy amounts to approximately two thirds of the income from the bio-energy plant, leading to a gross margin of 10%. Once the subsidy has been granted it will apply for ten years.

Applications for the subsidy can be made only once New Mixed Farm holds all the necessary licences and permits. If the experience with subsidy schemes in the Netherlands is anything to go by, whether the subsidy will still be there and in what form is uncertain. The future will need to demonstrate to what extent the bio-energy plant business model is sufficiently robust on this score.

New Mixed Farm's value proposition rests partly on benefits of scale. This aspect of the value proposition does however have its flipside in that certain elements in society are opposed to what they regard as factory-based animal production that is demeaning to animals. On account of the intensification of meat production, the innovative business model consequently also creates fresh opposition, even though the intensification is sustainable in nature.

4.6 From plan to investment

Even though New Mixed Farm has not yet been built, it is the most advanced agroproduction park in the Netherlands involving animal production.

To date the following concrete results have been achieved:

- The Municipal Executive and the Municipal Council of Horst aan de Maas have come down in favour of the establishment of New Mixed Farm in the municipality.
- The 'No to Large-Scale Livestock Farming' citizens' initiative has not succeeded in Limburg, in contrast to the situation in a number of other provinces.
- The B.V. Bio-energiecentrale Maashorst has been established for the construction of the bio-energy plant. The business plan for the plant is complete.
- The entrepreneurs remain firmly resolved to proceed with New Mixed Farm.
- The Environmental Impact Assessmen has been submitted, together with the applications for the environmental and building permits for phase 1.

Financing

Getting New Mixed Farm off the ground will require an investment by the various owners. For the poultry farm, part of the necessary funding will be obtained from the sale of the present four poultry farms owned by Kuijpers Kip B.V. The investment in the pig farm will be financed by means of an ordinary bank loan. The bio-energy plant calls for an investment of over twenty million euros. Part of this investment will be funded by the owners Kuijpers Kip B.V., Heideveld Beheer B.V., Houbesteyn Groep and Christiaens Engineering & Development B.V. in the bio-energy plant. The other part will be financed by a bank loan.

Acquisition of the land

The land required for the establishment of New Mixed Farm belongs at present to the Municipality of Horst aan de Maas. At the point at which the EIA and the supplementary investigations result in final approval by the municipal council, sale to the entrepreneurs will be possible.

4.7 The lessons for the entrepreneur

The entrepreneur as project developer

Initially, the plans for fully closing the loops within the New Mixed Farm were highly ambitious. The entrepreneurs found themselves in the driver's seat quite soon after the start. The joint working sessions held by the entrepreneurs provided a useful reality check on the ambitious plans. New Mixed Farm has a lengthy development time-frame. Over the years the political playing field has changed, as have attitudes towards intensive livestock farming. Aspects such as animal welfare and animal health have become more important. Public support for the plans and also the support at government level changed during the development period. The entrepreneurs in New Mixed Farm adjusted their plans accordingly. An independent sustainability scan conducted on behalf of the municipality was particularly helpful in this regard. This provided a point of reference for the development of their proposition by the entrepreneurs, in the sense that sustainable development is an ongoing process in which entrepreneurs must be prepared to adapt to changing needs and perceptions within, of course, certain (economic) limits.

The entrepreneur as coach

Right from the start of the project most of the parties concerned understood that the New Mixed Farm innovation required more experimental space. It was the former Minister of Agriculture, Cees Veerman, who provided New Mixed Farm in 2004 with an undertaking of separate status. To begin with this 'sanctuary' meant that things got off to a flying start. It quite quickly became clear however that the separate status was an empty political gesture without substance. The commitment was never converted into actual separate status. Similarly the designation of a location in an Agricultural Development Area ultimately proved not to have any added value. The designation of 'sanctuaries' for sustainable innovations – in this case in the form of first of all separate status and then the designation of space in an Agricultural Development Area – is only really of help to entrepreneurs if the government puts these instruments into effect.

New Mixed Farm is on a scale not yet seen in the Netherlands, involving linked-up businesses and a technology that is not yet proven. In the licensing procedure the entrepreneurs repeatedly ran up against the fact that the technology is assessed against the benchmark of 'best proven practices': something which new technologies are unable to comply with by definition.

This combination of factors meant that on account of the lack of actual experimental space, extra-legal requirements were imposed on the entrepreneurs.

The entrepreneur as strategist

The entrepreneurs are working on New Mixed Farm in terms of the vision that agriculture will only have a future in the Netherlands if it is able to contribute positively towards the public debate. The strategy they have selected in this regard is that of

'Sustainable Intensification'. They are working on methods of production that are sustainable and aimed at a bigger scale. It is vital for the new methods of production to gain public acceptance and appreciation.

The entrepreneurs in New Mixed Farm have experienced that this calls for new competencies. For this reason they have hired external communication expertise. They have also hired an architect in order to blend the project into the landscape as effectively as possible.

The protests that arose towards New Mixed Farm in mid-2008 are to a significant extent based on emotions, uncertainties concerning the effects and 'Not In My Back Yard' feelings. In response the entrepreneurs initially decided to provide more factual information. This approach led primarily to greater mistrust, along the lines of Shell's Brent Spar syndrome. It is not a matter of creating or exchanging more knowledge, but of understanding the perceptions of the various parties and coalitions.

The key players, including the entrepreneurs, would be advised continually to ask themselves how they should position themselves, what arguments stack up in a particular context and where areas of agreement are to be found.

The first steps in this direction were taken in the New Mixed Farm development period of New Mixed Farm described here. That was done by trial and error. One awkward factor in the beginning was that the consortium of entrepreneurs was not yet stable. It therefore took a relatively long time for the entrepreneurs to develop a shared story of their own. Not until one has one's own story is it possible to engage with local residents.

On the other hand, concrete plans leave little room for discussion and public participation. The separate story must therefore leave enough room for genuine involvement by others.

The entrepreneur as games-leader

As soon as it became clear which entrepreneurs wanted to put their shoulders to the wheel of New Mixed Farm they built up mutual confidence in joint working sessions. These also provided an opportunity for them to outline what their input would be and what their goals and expectations were. A joint mission to China also helped bring them closer together.

The entrepreneurs had and still have the initiative in the New Mixed Farm project. The perseverance, resolve and motivation of these three entrepreneurs are particularly pronounced. The unwavering support for the entrepreneurs and management of the network by the innovation broker KnowHouse were also important.

In recent years the New Mixed Farm entrepreneurs have been confronted by all sorts of unexpected obstacles and a lot of resistance. In the same way that great trees attract the wind, so businesses genuinely seeking to make a quantum leap are apt to run into greater resistance than those favouring a gradualist approach. Others would probably have thrown in the towel long before, but these entrepreneurs show no signs of giving up. This is not just on account of the anticipated financial results – 'If I'd been in it for the money I would have sold my farm long ago' – but 'Because I want to pass on something that is worthwhile'. It is primarily thanks to the entrepreneurs' fighting spirit that New Mixed Farm is where it is now, namely on the threshold of realisation. The commitment of these entrepreneurs can certainly not be faulted!

The entrepreneur as spider in the web

The entrepreneurs in the New Mixed Farm are aware that this complex innovation will only come about if the right connections are established with other parties. Since they lack the time and expertise for this themselves, they have outsourced this important task to the knowledge broker KnowHouse. At the same time there is a steering group that

lubricates the entrepreneurs' dealings with external networks when it comes to issues of decisive importance.

Under the *'Sustainable Intensification'* strategy, the generation of public support for a sustainable, intensive method of production is an important element. Ultimately the success will stand or fall on the will and discipline of the stakeholders to immerse themselves consistently in the images and perceptions of other parties, and on the basis of that knowledge pragmatically to explore the scope for agreements with parties who see things differently. Only by establishing a sense of connection does it become possible to work together on more sustainable livestock farming instead of attacking each other by argument and counterargument. Instruments such as force field analysis and stakeholder analysis can be helpful in this regard. In order to gain public support in the near future, it can be vital for the entrepreneurs to engage someone who is experienced in acting as a bridge-builder between societal organisations and the private sector.

Following initial scepticism on the part of the entrepreneurs, the cooperation between the entrepreneurs and researchers worked particularly well. The researchers made valuable contributions towards the development of New Mixed Farm. The most important success factors were for the researchers to pay sufficient attention to commercial aspects, to have a coordinator linking up the research disciplines, and for there to be regular feedback to the entrepreneurs of the research results.

The entrepreneur as winner

The main environmental gains can be made by linking up business loops and cycles. This also increases the mutual interdependence of the businesses, which constitutes a risk to the continuity of the operations. In terms of business management that is less sustainable. The New Mixed Farm entrepreneurs therefore elected to develop three side-by-side businesses, and not just one New Mixed Farm. This greatly reduced the number of linkages in relation to the original plans and made them more manageable.

The entrepreneurs drew up a joint business plan for the bio-energy plant, plus separate business plans for the poultry farm and the pig farm. The sustainability story of New Mixed Farm was underpinned by a sustainability scan conducted on behalf of the Municipality of Horst. The fact that there was an independent client acted as an advantage. In certain areas the entrepreneurs subsequently also commissioned a number of smaller sustainability analyses focusing for example on the carbon footprint.

The pig farmer does not see his pigs being marketed or positioned especially on the basis of sustainability. Given the present bulk nature of the market, his pigmeat operations are based around a low cost strategy.

In the case of the poultry branch of New Mixed Farm, the added value is increased by the integration of production with the slaughterhouse and meat processing. The poultry farmer is seeking to market the chicken meat with the aid of a supermarket chain house-brand based around the sustainability story. By organising a specific sales channel through the supermarket, the farmer is hoping to achieve a higher margin.

Completing the business case for the bio-energy plant will depend critically on whether or not the government subsidy is obtained. If it is, it will – as matters stand – be valid for a period of ten years. That is still a long way off but the entrepreneurs will need to work on an alternative financial underpinning for the business case in good time – something that forms an inherent factor in working with subsidies. After ten years the cost of depreciation and interest will be significantly lower and it is not inconceivable that energy prices will have risen, so that alternative energy generation such as that by the bio-energy plant will become more attractive.

4.8 The present challenges

Permits

The first challenge is to obtain all the necessary licences and permits, so that the land can be sold to the entrepreneurs. The researchers have no more than limited influence over this process: their sphere of influence extends to co-operating in the various surveys and providing the information sought.

From resistance among members of the public to resistance among consumers?

The extent to which local or national resistance can lead to a decision that is prejudicial to New Mixed Farm is fairly unpredictable and not something that the entrepreneurs themselves are easily able to influence. In the (undesirable) event that the present resistance at local community level spills over onto consumers buying products from New Mixed Farm, this could adversely affect the earnings model.

Radio silence versus public support

Since mid-2008 the entrepreneurs have deliberately opted for a strategy of radio silence. They want to work away quietly on New Mixed Farm and to avoid wherever possible being sucked into the wider, national debate about large-scale livestock farming. While that is understandable it does have the drawback of a decline in support among other stakeholders, such as the Ministry of Agriculture, Nature and Food Quality. It also makes it difficult for various parties to design and realise a sustainable livestock farm jointly. Sustainable livestock farming comes about only if stakeholders manage to transcend their differences, actively explore the ground they have in common and jointly set to work. The submission of the EIA in mid-2010 gave the entrepreneurs more room again to tackle the public relations side.

Building and then…

Assuming that the permits are obtained, construction can commence. The following considerations then arise:

- Is sufficient cash available when required for the various stages of construction?
- The internal pricing of the products to be traded among the New Mixed Farm Enterprises has still to be determined.
- The actual subsidisation of bio-energy plant products will need to be worked out.
- Is the USP of the sustainable chicken products from New Mixed Farm sufficiently robust in itself to be sold under a supermarket own-label with sustainable connotations and for sales contracts to be concluded? Or are adjustments required or is extra recognition in the form of a particular hallmark or other cooperation with the societal organisation desirable? *(by way of analogy to Rondeel case; see elsewhere in this book).*

THESE INNOVATIVE ENTREPRENEURS PULL OUT ALL THE STOPS

The entrepreneurs in New Mixed Farm (from top to bottom): Gert-Jan Vullings, Huub Vousten, Marcel Kuijpers and Martin Houben (Photo: Ruud Pothoven, KnowHouse B.V.)

SUSTAINABLE VALORISATION

Landmarkt, MijnBoer and Rondeel

The first Landmarkt gets off the ground beneath the haze of Amsterdam (Photo: Mugmedia, Wageningen)

5. LANDMARKT

covered farmers' market as metropolitan meeting place

5.1 The challenge

The most important purchasing criterion for retailers when putting together a fresh-produce range remains the purchase price. The product range is subject to the trend whereby more and more pre-packaged fresh products are being carried. Consumers consequently become alienated from the production and processing of their food. At the same time, particularly among urban consumers, there is also interest in traditionally produced food of good quality. Landmarkt meets this demand with the aid of a new retail formula. It is designed to become a chain of modern, covered marketplaces on the fringe of cities, specialising in daily fresh products sourced from farmers and traditional producers in the region. Producers, processors and Landmarkt bear joint responsibility for the success of the formula. Producers are no longer an exchangeable factor. By shortening the chain, direct contact is (once again) established between consumer and producer. This is a two-way process: consumers see and hear how their food is being produced, while farmers and processors hear directly what the customer thinks about their products and receive a higher payment for their product and efforts (but are also required to do something extra for that). The aim is to generate 70% of total turnover from fresh produce (in the 'fresh corners'). In the case of a supermarket the figure is 40-50%. The shops will be partly run by farmers and traditional food processors or by partnerships of such people.

5.2 How did the innovation come about?

Origins in idealism

The initiators of Landmarkt, *Jan Boone* and *Harm Jan van Dijk*, were guided by a kind of idealism when they developed their

concept in 2007. The aim was to offer the Dutch producer a realistic price again and to add value for the producer. They sought to effect a change in the hierarchical structure of the chain: greater sustainability is not readily achieved when retailers primarily interested in price and shelf-life are in the driving seat.

Since 2008, Harm Jan van Dijk has been involved full-time with the development of the company, as managing director. The initiators are familiar with the food industry and are convinced that a formula such as Landmarkt is the right idea at the right time. Both are therefore personally investing in setting up the chain. When fleshing out the concept it was assumed that a new marketing segment had arisen as a result of changing consumer requirements:

- *Individualisation*. The number of one-person households iin the Netherlands will rise from 2.5 to 3.1 million in 2020.
- *Time is scarce*. Time is the new scarce good in society.
- *Ageing*. The proportion of over 55s is set to rise from 27% to 37% in 2030.
- *Sustainability*. Consumers are demanding that the suppliers operate responsibly.
- *Regionalisation*. As a counterpart to globalisation the region is becoming ever more important.
- *Dematerialisation*. The emphasis is shifting from quantity to quality and from tangible to intangible.
- *Authenticity*. The surfeit of processed food gives raise to an increasing demand for 'real' products.

Project partners

Landmarkt, VU University Amsterdam (Athena Institute), Wageningen UR (LEI and Department of Environmental and Social Sciences) and TransForum.

TransForum project

2009-2010

- *Convenience*. Everyone goes short of time. Convenient solutions are becoming ever more important.
- *Enjoyment*. Convenience and healthiness are fine, but the food needs to be tasty as well.
- *Health*. Consumers increasingly wish to establish healthy eating patterns.
- *Good behaviour*. Taking account of nature, climate, fair trade and the local situation.

The concept

Landmarkt is a new formula of an open marketplace with catering facilities that is designed to link up the city and countryside. LandMarkt specialises in tasty, natural, daily-fresh products from the region. It offers a wide range of bread, meat, fish, dairy products and food and vegetables, and has a limited range of all other day-to-day requirements (including a jar of peanut butter!). Meal components, and ready meals and salads prepared on location, form part of the range on offer. The fresh-produce range is seasonal. The key elements are flavour, variety and the introduction of old and new varieties. Wherever possible Landmarkt misses out the intermediate trade in the food chain and provides farmers and horticulturalists with a direct and high-margin sales channel to the consumer. Landmarkt is essentially based on a culture of openness in its dealings ('*if you don't believe it, come and have a look*'), not in issuing hallmarks.

Landmarkt has an authentic butchery and bakery. Salads, fresh pasta and ice cream are all prepared on site. The open kitchen allows it to be seen how products are made, and is also possible for products to be tasted. You can have a meal there or take a freshly prepared meal away.

The retail trade and the catering industry are integrated, turning Landmarkt into a meeting place as well.

Harm Jan van Dijk:

'Landmarkt is a new, innovative shopping chain for everyday shopping where buying, tasting and learning about food all merge into each other.'

Diversity of products

Supermarkets tend to suffer from product standardisation. To take one example, there are several hundred apple varieties that can be cultivated in Western Europe, but nearly all the supermarkets confine themselves to just Fuji, Granny Smith and Golden Delicious. They do so because these varieties are suitable for long-distance transport and automated harvesting. Landmarkt wants to introduce a greater diversity of products, which is possible since the transportation is over short distances.

Franchise formula

Landmarkt is a centrally managed franchise organisation that generates its income mainly from franchise fees. The overarching Landmarkt organisation is responsible for marketing, product range management and coordination of the activities required for the Landmarkt shops to operate. Regional fresh produce partners (farmers and traditional processors) will preferably be both a supplier of and a franchise-holder in a Landmarkt. Supported by the Landmarkt organisation they run their own fresh-produce range in the shops, either on an individual basis or as part of a collective. This makes for the shortest lines in the chain, the freshest produce and the biggest margins for the producer.

The first few shops have been managed by Landmarkt itself. This also applies to the running of the non-fresh product range and the catering facilities. In future these activities will probably be contracted out to franchises. For the present, therefore, the Landmarkt management team remains responsible for the first shop. Their experience and knowledge of retailing make it possible for them to handle specific problems concerning the management of a Landmarkt shop themselves.

Accentuating the distinctiveness of Landmarkt

In order to take the Landmarkt concept further, a TransForum project was set up in 2009 together with Wageningen UR and the Athena Institute of VU University Amsterdam. Specific attention was paid in a number of working sessions to the question of: What is it that sets Landmarkt apart? The recommendations from the sessions were used to tighten up the purchasing policy. The result was that all Landmarkt products selected must, as a minimum, comply with such criteria as affordability, animal-friendliness and lack of unnecessary additives. Supplementary Landmarkt criteria are regionality, flavour and directness:

- Regionality: the producer will for preference be located no more than 50 km from the Landmarkt shop.
- Flavour: each product is assessed on taste.
- Directness: no unnecessary links in the chain. Consumers must be able to trace products so that they can tell where a product has come from and from which farm. If the farm is nearby it must be possible for consumers to pay a visit.

The TransForum project has made use of knowledge institutes as network brokers, while strategic working sessions were held with the monitor in order to keep the long-term goals clearly in mind. The results of consumer surveys have been used to respond to consumers' wishes more effectively.

Cooperating with municipalities is required in order to obtain permits

Through the location of its shops, on the edge between city and countryside, Landmarkt acts as a gateway between the two. Landmarkt outlets will therefore preferably be located on the fringe of the city where there is good access, near arterial road and rail links. Obtaining the right permits is a lengthy process: the present Spatial Planning Policy seeks to keep retail outlets outside of the core areas.

At the same time many civil servants are enthusiastic about the Landmarkt concept and want new types of regional food chains to be integrated into their municipalities. For Landmarkt to be issued with a building permit cogent arguments had to be found in order to support an extension of the zoning area.

Landmarkt has conducted extensive talks with various municipalities and now has a well-filled pipeline of future business locations.

Organising public support: mobilising local residents

During the period in which the first permit for Landmarkt was issued in Amsterdam, the managing director of the company attended a village council/public consultation evening. The session with people living in the vicinity of the future Landmarkt attracted an unexpectedly high turnout. To begin with the local residents expressed their worries about the level of activity and traffic that would be generated by the Landmarkt. By the end of the evening, much of the opposition had melted away in the face of Harm Jan van Dijk's persuasive arguments and enthusiastic account.

Selection of suppliers

In 2009 meetings were held with potential suppliers. The contacts with primary growers/farmers wanting to team up with Landmarkt proved much easier than had been anticipated. Partners were rapidly selected for meat, fish and bread, who sell their own part of the product range in the shop themselves. In the case of fruit, vegetables and cheese a different kind of partnership was pursued. It did not prove possible to come to an arrangement with groups of growers. Landmarkt therefore decided to introduce a second sales model. At the suggestion of the growers, this also involves a modification of the financial model: the margin on top of the purchase price and payment is divided 50/50 between Landmarkt and the grower/cheese supplier.

This means three different selling methods for Landmarkt:
1. On the basis of a franchise agreement. For products such as bread and meat. The producers sell their own products under a franchise agreement.
2. On the basis of a shared-risk contract. This applies especially to potatoes, vegetables and fruit. The growers sell their produce on the basis of the costs incurred and then share the profit margin with the Landmarkt initiators on a 50/50 basis.
3. On the basis of separate Landmarkt purchasing. Other products – such as pasta – are purchased by Landmarkt itself and sold for its own account.

2008	Jan Boone and Harm Jan van Dijk work jointly on the Landmarkt concept
2009	TransForum project with LEI Wageningen UR, Wageningen University and Athena Institute
	Consumer, chain and spatial planning research by Wageningen UR (including LEI)
	TransForum working sessions, among other things to determine Landmarkt core values
	Working sessions with specialists and societal organisations in order to determine the most important sustainability indicators
	Landmarkt management team complete
	Management team travel to USA and elsewhere and visit new retail formulas
	Credit crisis complicates financing: extensive consultations with potential investors
2010	Amsterdam permit and commencement of construction
2011	Opening of the first Landmarkt in Amsterdam

5.3 Key figures

- The supermarket chains in the Netherlands have total sales of around 29 billion euros per year. These rose in 2009 by 900 million.
- In ordinary supermarkets, 40-50% of sales are generated by fresh products: bread, meat and meat products, fish, fruit and vegetables, dairy and ready-meals.
- Dutch consumers spend an average of 1,070 euros a year on fresh produce.
- In addition they spend 560 euros a year on non-fresh products.
- The first Landmarkt shop opened stores in April 2011 and the aim is to have several dozen outlets in operation within 10 years.

5.4 The added value of Landmarkt

The first Landmarkt location

The first milestone was the opening of the first outlet (7 April 2011) in Amsterdam. This location was the first to obtain all the necessary construction permits. Strategically, Amsterdam is a good location. Building up a new market segment calls for a particular consumer profile, which can be found in and around the area selected.

Another advantage is that the opening of the first outlet was an important occasion for generating widespread media attention. This is easier to achieve in the biggest city in the country.

The advantages of Landmarkt

Landmarkt is a unique fresh-produce concept, in which the unique selling point is the combination of food, education and shopping.

The unique feature is that the Landmarkt formula concentrates on:

- Tasty and fresh regional products.
- Direct distribution from the producers to the point of sale (i.e. without any distribution centre).
- Working with suppliers on the basis of different contracts. Farmers or traditional suppliers may offer their wares under the Landmarkt roof either individually or as a partnership.
- The provision of detailed information about the provenance of products. Landmarkt links the farmer directly to the consumer by means of a verifiable and substantiated story.

Landmarkt

Landmarkt brings consumers and producers together through regionally produced, natural food. It is an innovative retail chain in which buying, tasting and learning about food are all intermingled.

⊕ The advantages

Regional products

Supply of tasty and fresh products derived from farmers and traditional producers in the region.

← max. 50km →

⊕ Promotion of sales by small entrepreneurs

⊕ Fair price

70% of the sales come from fresh products (compared with 40% in a mainstream supermarket)

Landmarkt

Covered marketplace where natural, affordable and animal-friendly products are on offer.

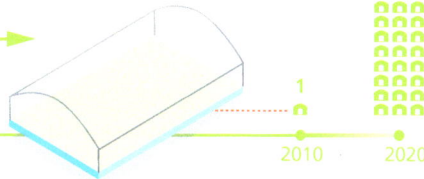

30

1

2010   2020

⊕ Clear product origin

⊕ Direct distribution from producer to the point-of-sale

Regional consumers

Consumers go to the Landmarkt in order to taste, buy and prepare regional products or to learn about food and health.

⊕ Consumers more food-aware

⊕ Tasty and fresh products from the region

⊕ More information on product origin

What does Landmarkt have to offer?

products on sale include:

fish   meat   fruit

bread   dairy   vegetables

Learning about food   Catering industry

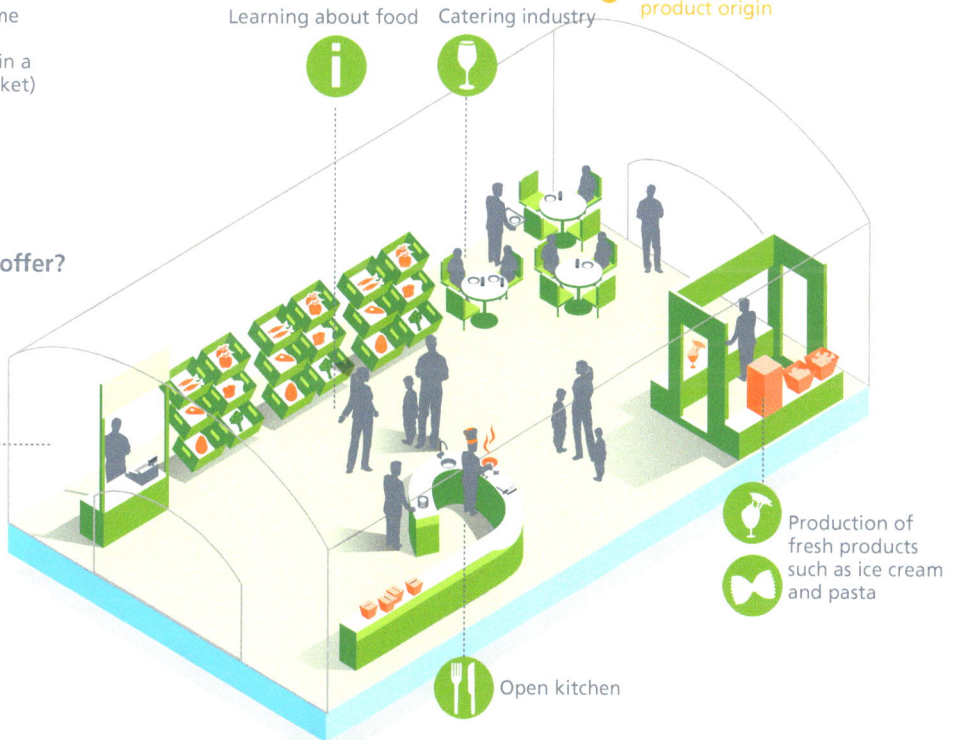

Production of fresh products such as ice cream and pasta

Open kitchen

The sustainability performances

People

- Generating a fair income for farmers and traditional processors of food, and giving them an equal position in the chain. This is made possible because they are fully-fledged partners who help decide on the allocation of the margin within the chain. In the traditional sales channel the farmers occupy a different position: they send their products to auction and then have no influence over the price paid for their products.

- The new allocation of margins means that it can become profitable to produce other (e.g. specifically local) products as well.

- Learning about food. Landmarkt wants to inspire consumers by showing how food is grown and processed. The catering facilities with a large, open-plan kitchen will encourage consumers to make different, more aware choices. Landmarkt wants to develop recipes in order to encourage healthy eating. For the consumer it is important to know that the chain is transparent (i.e. to know where the products come from). The education of consumers and also schoolchildren will be promoted by activities concerning health, cooking, preparation and tasting and by visits to suppliers in the region. Landmarkt wants to develop educational programmes in collaboration with schools, under which schoolchildren visit producers. It is also the intention for consumers to be able to take part in workshops in the shop.

- Landmarkt invests in a green environment for the shop, for example by setting up a production garden around the shop. Landmarkt is making a structural contribution to the environment by taking account of the management, preservation and accessibility of the rural hinterland.

Planet

- Regional products: the sale of products produced in the same region results in less transport and hence lower CO_2 emissions.

- Each product is assessed in terms of sustainability performance. In the case of meat for example, animal welfare, transport and slaughter are evaluated, while in the case of tomatoes the emphasis is on the use of water, fertiliser and energy.

- Seasonal products. Seasonal products are produced in optimal ecological circumstances. Landmarkt intends to inform customers more effectively in this area, thereby promoting seasonal produce.

Profit

- Fair price. A short chain and shared responsibility for the product make it possible to distribute the margin differently within the chain. The short chain means that products can be sold at a reasonable price while also paying producers a fair price ('cost price plus'). This can be achieved by increasing the added value provided by the farmer (in terms of production, transport and marketing) or by linking the consumer price to production costs.

- Promotion of sales by small entrepreneurs: numerous small entrepreneurs each specialising in a specific and attractive product lack the capacity to market their product themselves. They are too small to produce the large volumes required for supermarket chains.

- Gives these entrepreneurs a chance to sell their products at an attractive margin.

- New market segment: particularly in the big cities more and more consumers do not want to cook every day. Landmarkt offers tasty products for these customers, as well as takeaway meals prepared on the spot.

- Fresh produce: the Landmarkt formula provides for a reduction in the delivery time between harvesting and sale (down from 7-8 days to just 1-2 days). Particularly in the case of fruit and vegetables this significantly improves the flavour.

- Major investments have been made in order to set up the Landmarkt formula. The key question is whether these large investments can also be recouped in time.

SWOT analysis of the sustainability performances

Strengths

- A positive effect on producer incomes.
- Psychological benefits for employees as they work in an inspiring setting and the pay is better than in mainstream supermarkets.
- Community involvement (through educational programmes).
- The presentation of healthier products (such as bread with a lower salt content).
- The Landmarkt concept can be scaled up straightforwardly.

- The quality of the local environment is given a boost since suppliers/farmers are encouraged to be active in meadow-bird management and nature conservation.

Weakness

- The investment required to set up a totally new concept including the production chain is high and consequently involves a high risk.

Opportunities

- Developing activities whereby consumers take responsibility themselves for more sustainable production (e.g. by organising a weeding or harvesting day).

Threats/risks

- Sufficient availability of agricultural products in the region.
- Striking a balance between a fair price for producers and an acceptable price for consumers.

5.5 The value creation model

Notes on the value creation model

The organisation of a new type of chain cooperation has made it possible to distribute the margins in the chain differently. Primary producers and processors are paid a different price. Added value is organised by responding to consumers' wishes regarding quality, flavour, transparency, regionality and sustainability.

Together with suppliers Landmarkt has set up an organisation offering regional quality products in a covered marketplace, while at the same time providing catering facilities and educational opportunities. The new business model also allows producers to become co-shareholders in the concept.

The creation of values may be shown as follows in the model (see next page).

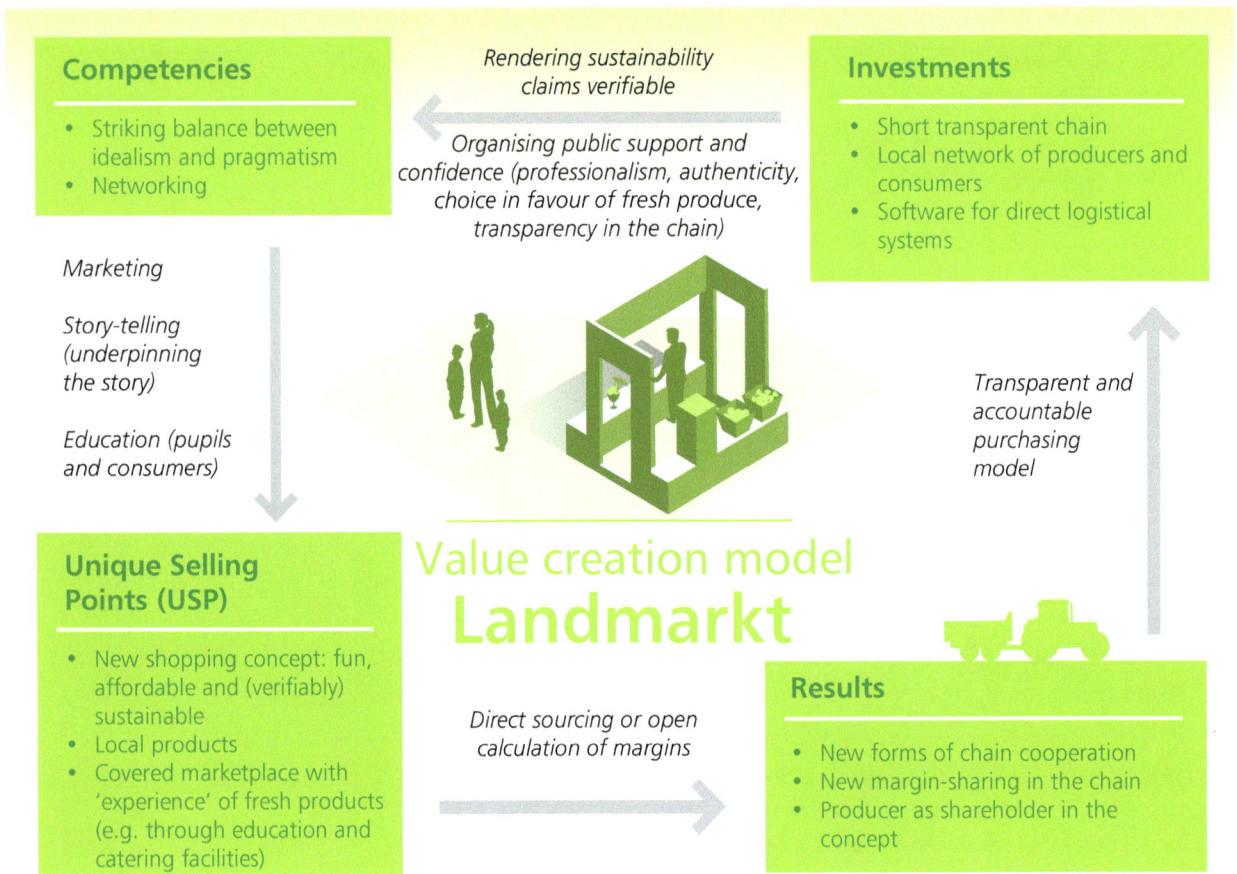

Competencies

- Striking balance between idealism and pragmatism
- Networking

Rendering sustainability claims verifiable

Organising public support and confidence (professionalism, authenticity, choice in favour of fresh produce, transparency in the chain)

Investments

- Short transparent chain
- Local network of producers and consumers
- Software for direct logistical systems

Marketing

Story-telling (underpinning the story)

Education (pupils and consumers)

Transparent and accountable purchasing model

Unique Selling Points (USP)

- New shopping concept: fun, affordable and (verifiably) sustainable
- Local products
- Covered marketplace with 'experience' of fresh products (e.g. through education and catering facilities)

Value creation model
Landmarkt

Direct sourcing or open calculation of margins

Results

- New forms of chain cooperation
- New margin-sharing in the chain
- Producer as shareholder in the concept

5.6 From plan to investment

The Landmarkt investment plans and business model

The average supermarket in the Netherlands has an annual turnover of 4.9 million euros. This is achieved with an average floorspace of 800 m². Landmarkt wants to generate higher sales per outlet, particularly by supplying more products with a higher added value, such as takeaway meals and catering facility products. The shops have been set up more spaciously. The estimated turnover per m² is comparable. The Landmarkt business model results in a different allocation of value, since the direct sales mean that Landmarkt does not use a distribution centre.

Attracting investors

An initial investment of several million euros was needed for the development of the Landmarkt chain. This money is being used to build the first few shops and the cash flow can be generated to finance further outlets.

The market situation for risk capital was not favourable in the period 2009-2010, and it accordingly took a lot of time and energy to put together the necessary capital. This was done in three ways:

(a) The initiators and the management team made private investments. As co-investors, all of them are also shareholders in the company.

(b) The 'LM investors' fund has been established for smaller scale investments. Interested investors are given the opportunity to participate jointly in Landmarkt Holding. The minimum subscription is 100,000 euros. Landmarkt had a brochure printed for this fund and held meetings for potentially interested investors. LM investors has 20 participants who are sharing in the operating result.

(c) Bigger investors and private equity companies have been attracted.

5.7 The lessons for the entrepreneur

The entrepreneur as project developer

Landmarkt wants to promote sustainability through making a change in the hierarchical relationships within the chain. In the Landmarkt food chain the producer and Landmarkt share responsibility for the quality of the product and both parties share in the profits. Investments in taste, product innovation and sustainability are visible to the consumer, so that the added value can be recouped. Empowerment of the grower is central for Landmarkt, so that a good and tasty product becomes available for the consumer.

The entrepreneur as coach

The innovation process was inspired and developed by a number of committed entrepreneurs. Without their enthusiasm and perseverance, Landmarkt could not have got off the ground. Commitment on the part of the entrepreneur is vitally important.

The entrepreneur as strategist

Landmarkt has decided to emphasise the link with the region in its positioning strategy. A direct link with the region gives consumers the feeling of freshness, better flavour and more direct contact with the producer. Regional products also allow Landmarkt to tell the story behind the product more readily. Although sustainability is a guiding concept for the management in its procurement policy, it was deliberately decided not to position Landmarkt as sustainable vis-a-vis consumers.

Sustainability is a very broad concept and in positioning a sustainable concept there is a risk that elements of the business that are not so sustainable will become exposed to criticism. The most important consideration, however, is that positioning on the basis of sustainability does not generate any competitive advantage. If doing so works, competitors can then also start making sustainability claims, whereupon the consumer is no longer capable of drawing a distinction. Landmarkt therefore prefers to position itself as a formula for regional products, meaning that the products are fresh and tasty. This is not readily imitated by competitors, as this would require a different business model.

Harm Jan van Dijk, managing director of Landmarkt:

'Landmarkt makes good and tasty food with a clear provenance available to a large group of consumers.

These will be modern, sparkling shops, in which account is taken of the cultural history and harmonisation with the natural environment.

The Landmarkt shops will where possible provide a view of the environs.'

The entrepreneur as games-leader

Commitment on the part of the entrepreneur is vitally important. The innovation process was inspired and developed by a number of committed entrepreneurs. Without their enthusiasm and perseverance, Landmarkt could not have got off the ground.

The entrepreneur as spider in the web

In order to set up a robust Landmarkt organisation a broad base of support is required.
Three years before the opening of the first shop talks were entered into with:

- Numerous municipalities: for a permit to set up a retail outlet on the boundary between city and countryside.
- Banks and investment companies: for financing.
- Potential suppliers: for concept development and contracts.
- Knowledge institutes: for knowledge concerning new concepts and consumer behaviour.
- Other innovative concepts: for inspiration.
- NGOs and independent advisers: for determining objective sustainability indicators that were used in order to redefine purchasing policy.
- Internal: a conscious effort was made to set up a relatively heavyweight management team.

The entrepreneur as winner

The ambition is to grow into a chain of several dozen outlets over the next ten years. A concept such as Landmarkt can only be realised by scaling up, and Landmarkt has taken account of upscaling from the start. During the planning phase, Landmarkt invested heavily in setting up a large network. Extensive consultations with various municipalities mean that there is now a well-filled portfolio of potential locations.

5.8 The present challenges

Over the next few years explicit attention will need to be paid to:

- Customer loyalty. It is not difficult to induce a consumer to make a single purchase. The real challenge is to generate repeat custom.
- Building a network of customers. Customers must have the feeling that they have a real influence on the way in which their food is produced.
- Spreading customers over the week. Supermarkets traditionally see a huge peak in sales at the weekend and lower sales during the week. On account of the freshness of the product – especially in the case of takeaway meals – there is a particular need to have enough customers every day.

The mission
'Landmarkt is the preferred daily purchasing and food chain for the aware modern consumer and producer'

The ambitions
- *Making sustainably produced tasty food, with a clear provenance, available to all consumers*
- *The creation of a transparent, local-for-local, food chain*
- *Generating a future-proof income for farmers and traditional processors of food, and giving them an equivalent position in the chain.*

MODERN MARKETPLACE WHERE THE CONSUMER MEETS THE PRODUCER

Harm Jan van Dijk, managing director of Landmarkt, near the entrance to the first Landmarkt location (Photo: Mugmedia, Wageningen)

The innovative shop concept in which Marqt collaborates with MijnBoer started in 2007 in Amsterdam (Photo: Mugmedia, Wageningen)

6. MIJNBOER

healthy and tasty food for all

6.1 The challenge

The market for potatoes, fruit and vegetables has largely developed into a commodity market. Particularly on account of the concentration of the wholesale and retail trades and the large-scale imports of fruit and vegetables from Brazil, India and elsewhere, the pricing pressure on fruit and vegetable products has increased. Supermarkets are largely guided by price in their purchasing policy. Their relationships with producers depend primarily on the product prices. It is therefore difficult for producers to invest in extras, such as a sustainable method of production, good flavour and quality, as this generates virtually no return in the traditional sales channel. Sales of fresh produce in traditional chains are largely driven by considerations of scale, uniformity and associated cost reductions.

MijnBoer ('MyFarmer') wants to make a shift in favour of sustainability, not just with the introduction of a hallmark but also by a change in the hierarchical relationships within the chain: greater sustainability is not possible when retailers are in the driving seat and solely emphasise price and keeping properties.

MijnBoer is an initiative on the part of fruit and vegetable growers who are seeking a strategy of differentiation so as to obtain a fair price for their products. This differentiation strategy is based in the first place on (a) the sale of more sustainable, tastier fresh products and (b) a more varied range of products.

In order to achieve this a direct link was established between consumers and growers. A direct link calls for a different kind of cooperation within the chain: a chain based on transparency. This involves telling the story behind the products. The fact that the farmer is paid a better price in the MijnBoer food chain makes it possible for producers to opt for a new style of farm management, based around offering quality products at a better price and more sustainable business operations.

MijnBoer helps growers obtain certification from the environmental hallmark association Milieukeur. Another sustainability aspect that has always been a prominent feature within MijnBoer is that of cutting down the amount of wastage (in the sense of unsold goods). The basic principle is that ultra-fresh products only should appear on the shelf or be used in salads.

The vision of MijnBoer
'Making healthy and tasty food available for all.'

6.2 How did the innovation come about?

Value proposition

The initial ideas for the establishment of a new type of regional fresh produce chain arose in 2005 in a collaborative venture between two organisations, *Ruraal Park* and *Foundation de Groene Hoed*. Ruraal Park was a developer of retail and experience concepts, and Groene Hoed an association of producers seeking to bring the city and rural areas closer together. Groene Hoed was at that time a supplier of regional products for the Amsterdam catering industry.

The idea of setting up a regional fresh produce chain took firmer shape in 2006. The two foundations are based on differing underlying principles that reinforced one another during the predevelopment process. A merger between Groene Hoed and Ruraal Park in 2007 resulted in the incorporation of Ruraal Park West B.V., trading under the brand name MijnBoer B.V.

Project partners
Buck Consultants International, MijnBoer B.V., VU University Amsterdam (Athena Institute), Wageningen UR (F&BR) and TransForum.

TransForum project
2007-2010

MijnBoer seeks to supply the urban market with tasty, seasonal products. The name 'MijnBoer' ('My Farmer') relates to the direct link between consumers and producers. Direct from the farmer means in the case of MijnBoer short supply lines and as few intermediate links as possible.

The formation of MijnBoer led to active involvement with the commercial side of the concept. In order to take the concept forward, a TransForum project was set up in 2007 together with Wageningen University and the Athena Institute. The project was aimed at knowledge development and the encouragement of a learning process within the project.

Mutual dependency within the chain

The first sales channel for MijnBoer was Marqt. Marqt got off the ground in 2007 as an innovative retail concept offering quality products straight from the producer. The first retail outlet opened its doors in Amsterdam in February 2008, followed by a second in Haarlem in August 2008.

Under a sale or return contract, MijnBoer delivered a complete range of fruit and vegetable products. The farmers continued to own the products, even once these were on the supermarket shelf. Producers were consequently responsible for the choice of products, a reduction in wastage, and product renewal. In exchange Marqt received a margin on all goods sold.

In contrast to the normal practice in conventional retailing, the contract and negotiations between MijnBoer and Marqt were constructive, being based on mutual dependence, good sales figures and certainly also trust.

Quirijn Bolle, owner of Marqt:
'Our shop is in no matter comparable with other supermarkets.'
Marco Duineveld, managing director of MijnBoer:
'And nor can our farmers with traditional farmers.'

The loading of a brand

2008 and 2009 were used in order to load the 'MijnBoer' brand with four important criteria: flavour, transparency, sustainability and quality. All the products in the range must satisfy these four criteria.

MijnBoer helps growers work towards certification from the environmental hallmark association Milieukeur. Products should wherever possible be grown in the ground without the use of pesticides. The preference is for Milieukeur-certified or organically grown products. Surprising, sustainable seasonal products also form part of the product range. The price paid for the products must be fair (i.e. cost price plus). The products must above all be fresh and tasty, i.e. they must have good flavour but also pay attention to taste perception. Suppliers are closely involved in the way their products are offered to the consumer.

Under the MijnBoer formula, transparency means a direct relationship with the grower. This also applies to transparency between elements of the chain from producer to consumer. Transparency not only helps ensure quality throughout the chain and the utilisation of suppliers' know-how, it ultimately also helps obtain commercial value from quality.

The consumer is often required to make a visual decision about whether or not to buy a product. MijnBoer products must be associated with quality criteria such as sustainability and flavour – criteria that are not often immediately apparent. MijnBoer organises this by creating a brand, by involving suppliers in its category management and by creating transparency between the various links in the chain.

Product innovations made possible by an increase in volume

MijnBoer also offers varieties that are not sold through traditional outlets, thereby increasing the diversity of the product range. MijnBoer has for example also introduced new and tasty strains. One example is apples that are not sold through traditional retail outlets. There are more than 500 different kinds of apples and just 20 varieties are sold in traditional supermarkets.

The development of product innovations together with

producers proved possible only through the combination of partnerships with producers and the organisation of a critical mass in terms of volume. Trust and economic certainty allow a different mechanism to come into play. Volume came to be seen as an important precondition for realising the ambition of a more diverse product range.

New partners

MijnBoer developed a network organisation, in which the producer and MijnBoer became closely bound up with one another through the exchange of activities. In 2009 other retail channels apart from Marqt were added to the clientele. MijnBoer's sales rose with deliveries to the restaurant chain La Place restaurants and food service company Vitam Catering. Both were anxious to set themselves apart by offering healthy, varied and local produce. In this regard Vitam opted deliberately for a strategy of differentiation and, together with MijnBoer, provided information on the origin of products and introduced unusual local products.

MijnBoer organised the contacts with farmers and customers and so was responsible for the coordination for all the partners. In exchange it received a margin on all goods sold. A characteristic feature was that MijnBoer entered into partnerships with customers and farmers. Both were closely involved in the development of new concepts. A good relationship with both supplier and customer was regarded as essential for the further development of the company. In the case of Vitam this led to a very strong relationship. MijnBoer helped draft the Vitam mission statement, was present at acquisition meetings and developed a new concept with Vitam, 'the Fresh Produce Market'. The Fresh Produce Market is a stand-alone outlet within the Vitam range offering both fruit and vegetables and MijnBoer groceries. Images of producers and products are shown in a slide presentation above the fresh market. The development of the Fresh Produce Market has given a new boost to producers' aspirations to provide the grower with a face and to restore the relationship between consumer and producer. The first Marqt store opened its doors in Amsterdam on 10 March 2010. Vitam has an exclusivity right to the new concept for two years.

De Kwekerij

De Kwekerij ('the Market Garden'), a joint venture between MijnBoer and Landzijde Foundation, was opened in March 2009. De Kwekerij employs people dependent on care. The products they harvest are marketed via MijnBoer. For MijnBoer this was a deliberate choice: by cooperating with De Kwekerij, MijnBoer wants to demonstrate that as a trading partner it is close to the soil, in contrast to the average trading partner, which acts in isolation from the cultivation process. MijnBoer wants to encourage other producers to use the experience gained with De Kwekerij in the cultivation of special products.

MijnBoer International

Consumers are not used to eating purely regional food. The MijnBoer customers sought a total range of products and year-round availability, including pineapples and bananas. MijnBoer therefore links up regional sourcing to international sourcing, taking into consideration when certain products can and cannot be regionally sourced. Important factors in this regard are quality and taste.

Similarly in the case of international sourcing the aim was to create the shortest possible chains, with the maximum provision of information from the producer for the consumer. The grower must receive a fair price for his products, the chain must be transparent and the product must be tasty and of high quality. MijnBoer does not make use of air cargo. This makes its international sourcing more sustainable than that in the mainstream trading sector.

Further increase in volume

2010 saw a further increase in sales with the provisioning of all one hundred Vitam locations. Whereas in the initial period the decision was to supply a wide product range with all sorts of weight variations, it was now decided to supply a limited number of specific weight categories. A change was also made to the product range: whereas MijnBoer had initially focused on more unique products, the brand concentrated much more in 2010 on regular, fast-turnover products such as cucumbers and tomatoes.

Marco Duineveld (right), founder of MijnBoer B.V. (Photo: MijnBoer)

MijnBoer does however try to preserve a distinguishable choice, for example by supplying not just ordinary tomatoes but also special products, such as a chocolate cherry – a type of wild tomato.

In determining the product range, a trade-off is constantly made between unique/special and the affordability of the product. Sales to La Place also continued to grow with the same products as those supplied to Vitam. This generated high logistical and administrative efficiency.

Changeover from specialty products to the mainstream channel

The collaboration with various customers has created new logistical flows. In the initial stage, MijnBoer handled the logistics for Vitam and La Place itself. This rapidly proved to be less than ideal, in that coverage of the entire distribution area called for a professional logistics service-provider. MijnBoer sought a suitable partner, the most important criterion being that the partner should also be able to help achieve higher sales and an increased volume. With an increase in volume MijnBoer can become significant for its producers. Only at this point is joint development possible, as investment in product development can then be recouped.

One of MijnBoer's ambitions remains that of bringing back product diversity. This can be achieved by introducing new products or reinstating 'forgotten' products in the range.

In June 2010 MijnBoer merged with wholesale company Sligro/Smeding. This meant that the MijnBoer concept became incorporated into the mainstream channel. Within Sligro, MijnBoer has become the sustainable line of fresh products, first of all for the foodservice channel and in 2011 also for retailing.

MijnBoer will then provide the input for the 'Honest and Delicious' brand. This label tells the story behind the products. In the shops, on packaging and on the Internet, the label allows customers to select products on the basis of such criteria as organic, fair trade, sustainability and regionality. The 'Honest and Delicious' brand is designed to communicate clearly to consumers in their supermarkets and to food-service customers. The shelf in the supermarket and the Internet become important factors for guaranteeing the 'Honest and Delicious' concept.

6.3 Key figures

- Total sales of fruit and vegetable produced in the Netherlands are estimated at 2.4 billion euros (on the basis of production prices), resulting in sales of approximately 10 billion euros (consumption prices).
- This production is divided into 441 million euros for fruit and 1988 million euros for vegetables.
- The sales of fruit and vegetables are made through supermarkets (82%), markets (8%), vegetable shops (5%) and other (5%).
- Consumer spending on organic food increased by 11% in 2009 from € 583 to € 647 million.

2005	Ruraal Park and Groene Hoed work on the concept of a regional fresh-produce chain
2007	Start of TransForum project together with Wageningen University and Athena Institute
	Commencement of collaboration with Marqt
	Formation of RuraalPark West B.V., trading under the name MijnBoer B.V.
2008	Opening of first Marqt outlet, delivery of fruit and vegetables on sale and return basis by MijnBoer
	Commencement of deliveries to Vitam locations (20 locations)
2009	Commencement of deliveries to La Place locations (15 locations)
	Cooperation with Landzijde and opening of De Kwekerij in Osdorp
2010	Opening of third Marqt outlet
	Provisioning of 100 Vitam locations and all La Place restaurants
	Opening of first fresh produce market within Vitam
	Takeover by Sligro/Smeding

MijnBoer

MijnBoer produces healthy and tasty food for all and brings the producer and consumer together. How does this work?

➕ The advantages

Producer

The farmers sell their products via MijnBoer or directly to retail chains.

MijnBoer

MijnBoer provides a direct link between producer and consumer.

Shops, catering industry

Fresh MijnBoer products are on the shop shelf.

1-2 days

MIJNBOER

FRESH

7-8 days conventional vegetable transportation

- ➕ Receives a good price from MijnBoer through the branding of his products
- ➕ Works with MijnBoer on quality improvement, sustainability and productive innovation
- ➕ MijnBoer develops new concepts together with producers and customers

- ➕ No air freight
- ➕ Less environmental pollution in production and no use of air freight
- ➕ Environmental certification

Foreign products

- ➕ Fresh seasonal and tasty products
- ➕ Large volume makes greater product range possible
- ➕ Story about the product and clear product origin
- ➕ Less wastage from storage requirements, + processing in salads

La Place
Marqt
Sligro/Smeding
Vitam

MIJNBOER

MIJNBOER

FRESH

sales of products such as:

fruit   vegetables

6.4 The added value of MijnBoer

The sustainability performances

People

- Consumers attach value to tasty food and fresh products. MijnBoer meets consumers' demand for fresh, tasty products.
- By offering familiar, forgotten and also new quality products, MijnBoer challenges the consumer to eat tasty (as well as healthy).
- Particularly in its partnership with Marqt, MijnBoer has placed the producers on the map: as the core organisation, Marqt has often stated in the media that it buys directly from the farm.

Planet

- Reduction in the percentage of wastage. The basic principle is to carry only ultra-fresh products on the shelf and to use products near their sell-by date in salads.
- The selection criteria for MijnBoer are the environmental hallmark Milieukeur or organic production, i.e. use of fewer pesticides in cultivation and no additives in processing.
- MijnBoer sells fresh products whenever possible, allowing for the fact that the quality of the product declines the longer it is kept in refrigeration.
- Products are imported only if it is impossible to source a comparable product in the Netherlands. Products are not imported by air, as airfreight is one of the most polluting ways to transport food and vegetable products.

Profit

- Branding is applied on only a limited scale in the fruit and vegetable sector. MijnBoer makes use of the various methods for establishing a direct link between producers and consumers. This encourages consumer awareness of their purchasing behaviour and provides important feedback.
- MijnBoer pays a better price to the farmer. This makes it possible for producers to place their operations on a different footing.
- In doing so, MijnBoer is reversing a trend. The need that farmers feel to scale up on account of low prices can be overcome by offering a tasty, high quality product at a better price.
- It is important to strike a balance between the price paid to the farmer and that paid by the consumer for the product.
- Sharing of the financial risks: agreements have been reached in mutual consultation in various MijnBoer supply chains, taking account of the joint interest.
- Long-term partnerships: the relationships in the MijnBoer supply chain also include agreements for annually guaranteed sales and/or joint development of innovations and marketing strategies among producers, MijnBoer and customers.
- Large (logistical) investments are needed in order to set up the MijnBoer chain.
- MijnBoer aims at a financially healthy formula, for which a certain turnover is required.

SWOT analysis of the sustainability performances

Strengths

- A positive effect on farmers' profitability.
- The creation of added value for agricultural products.

Weakness

- The investments that must be made in order to set up a new type of retail trade are high in comparison with mainstream retailing.
- A breakthrough such as that aspired to by MijnBoer is hard to measure.

Opportunities

- The encouragement of suppliers to invest in meadow bird management, more frequent crop rotation and other aspects of nature conservation.
- Organisation of visiting opportunities for primary school pupils to learn more about agricultural food production.

Threats/risks

- Striking a balance between paying a fair price and the higher price passed on to the consumer.

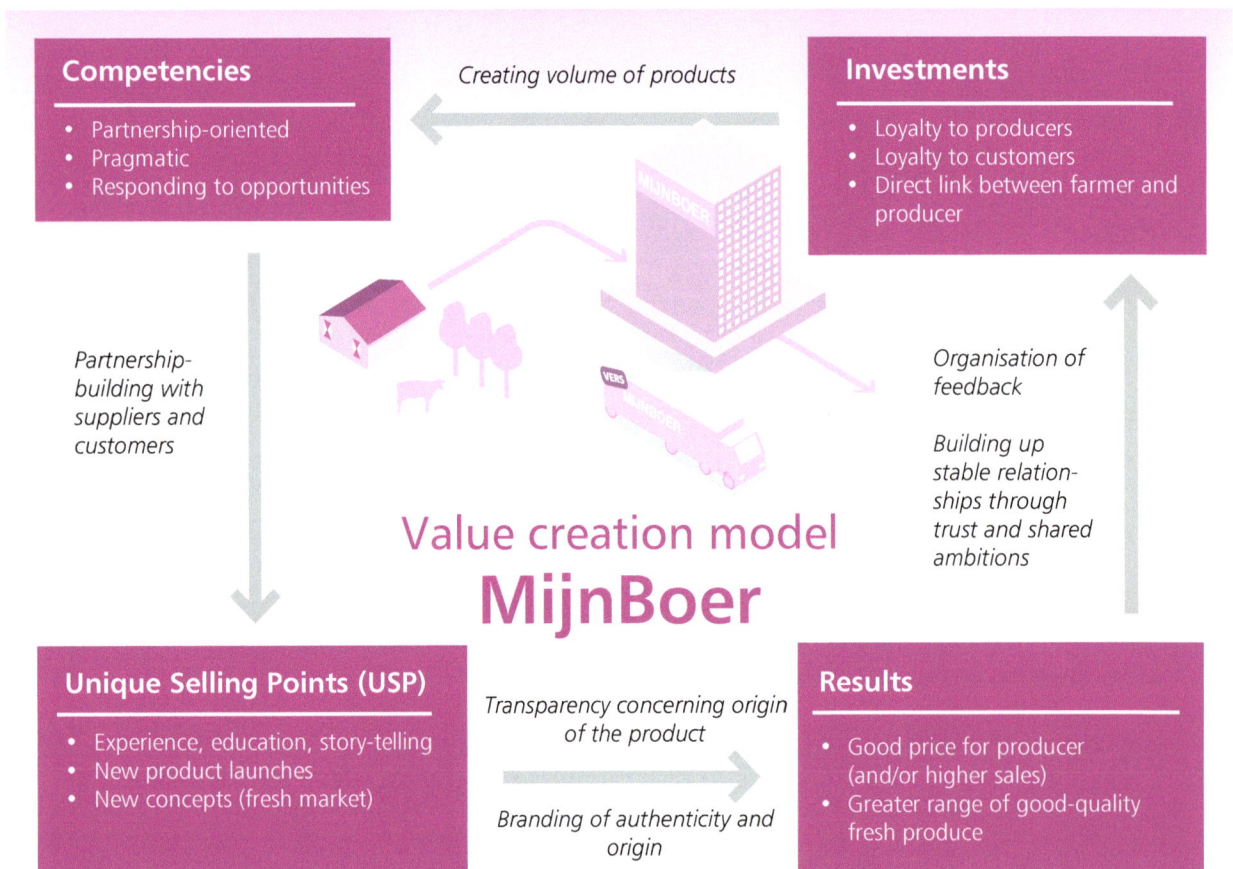

Competencies

- Partnership-oriented
- Pragmatic
- Responding to opportunities

Creating volume of products

Investments

- Loyalty to producers
- Loyalty to customers
- Direct link between farmer and producer

Partnership-building with suppliers and customers

Organisation of feedback

Building up stable relationships through trust and shared ambitions

Value creation model
MijnBoer

Unique Selling Points (USP)

- Experience, education, story-telling
- New product launches
- New concepts (fresh market)

Transparency concerning origin of the product

Branding of authenticity and origin

Results

- Good price for producer (and/or higher sales)
- Greater range of good-quality fresh produce

6.5 The value creation model

Notes on the value creation model

MijnBoer has evolved into a service organisation offering high-quality, tasty products. MijnBoer provides the producer and customers with knowledge and process management and support concerning product and information flows. It also provides producers and retailers with support so that the consumer can be offered variety, quality and flavour. The organisation of joint activities creates confidence and enables unique products and concepts to be developed. This makes it possible to enhance the MijnBoer brand, thereby allowing producers to be paid a fair price.

The creation of values may be shown as follows in the model, as shown alongside.

6.6 From plan to investment

MijnBoer is a service-provider, for which reason there are no large initial investments. The initiators have invested their own money. In addition the customers also invest in MijnBoer, (a) by agreeing to a 6% higher margin for MijnBoer (Marqt) and (b) through the provision of working capital (Vitam Catering). Office costs are kept low, and in order to keep the costs in the start-up phase down only two people were employed. The costs of these persons were paid in part out of the gradually rising sales.

MijnBoer's sales are now in the order of several million euros.

Earning money from sustainable products

The most important question is to what extent value can be created out of sustainability principles. The strategy was to have below-average costs with the short-distance transport, briefer period of storage and shorter chains. Leaving out the auction link results in lower CO_2-emissions (with the reduced transport) and good prices for producers. In practice transport costs turned out to be high when the logistics were organised on a small scale.

6.7 The lessons for the entrepreneur

The entrepreneur as project developer

MijnBoer's ambition was based on reversing a number of trends in the Dutch agrofood chain, such as the increase in scale in agriculture, a narrowing of the fresh produce range and the emphasis on price as the guiding mechanism for the payment of producers. For MijnBoer these were reasons to pursue the following aims:

1. Restoring of the relationship between city and countryside.
2. Shortening the chain by fostering cooperation among chain partners.
3. A focus on regional sustainable production.
4. Reducing product differentiation in the shop.
5. Increasing the quality and freshness of products.
6. Margin improvement for the producer.
7. Attention to food as an experience.

> **The ambitions of MijnBoer**
> *'To obtain a better price for the farmer by offering tasty, high-quality products by means of a shortened supply chain and the creation of involvement by the consumer in the product and the farmer.'*

In retrospect both the number and ambitiousness of the aims stand out. Three of the seven aims call for a sizeable innovation, namely direct contact between town and countryside, a shortening of the chain and a better price for the farmer. The lesson is that it is not possible for all the ambitions to be realised in a single step. A long-term vision and a step-by-step approach are required.

The entrepreneur as strategist

MijnBoer worked very pragmatically, but from a clear strategy. The strategy followed by MijnBoer has been one of *'Sustainable Valorisation'*.

MijnBoer was aware of the complexity of its ambitions. Achieving these required change on the part of a number of different stakeholders (farmers, logistics service-providers and shopkeepers). By means of a step-by-step approach MijnBoer managed to transferring its ideals of a specialty market into the mainstream retail channel.

The entrepreneur as games-leader

Agreements were reached concerning shared expectations. The innovation process began with the formulation of a shared ambition. It was notable that all the project participants had a clear, realistic and widely shared dream. The project participants consciously decided to work towards the realisation of that dream. Core values cited by MijnBoer are: transparency, honesty, responsibility, pleasure/fun and purity. In addition the dream conveys enthusiasm and energy. One success factor and, at the same time, learning moment concerning the project dream is: *'We meet each other in the what, and together seek the how.'* Following a lively discussion concerning the differences and similarities of view, the conclusion was drawn that the challenges that had been formulated applied primarily to the farmers. This observation led to a reorientation of the project activities, with the focus being placed more on the consumer.

The entrepreneur as spider in the web

The cooperation with new partners proved to be the key moments and meant that the ambitions remained more than just dreams:

1. **Ruraal Park and Groene Hoed**. The cooperation between the Ruraal Park and Groene Hoed organisations provided the means for putting the idea of a regional fresh produce chain into practice.
2. **TransForum**. External recognition and funding were received upon starting up the TransForum project. The ambitions were laid down on paper.
3. **Marqt**. Marqt is the prime MijnBoer sales channel, and serves as an example for the further relevant of the concept.
4. **Cooperation with producers.** The development of product innovations together with producers proved possible only through the combination of partnerships with producers and the organisation of critical mass in terms of volume. Trust and economic certainty allow a totally different mechanism to come into play.
5. **Vitam and La Place**. The provisioning of Vitam meant that MijnBoer was no longer dependent on a single sales channel. On the choice of an established partner it was said: *"You can't enter into an adventure in ten different places with a retailer who is also new. You therefore need to strike a balance between stable sales and new opportunities."*
6. **Sligro/Smeding**. A partner making it possible for MijnBoer products to be included in the mainstream.

Marco Duineveld:
'Once the entire organisation is based around Marqt, I will do what I can to pass on knowledge and experience and use this to take the next step of upscaling to a national system, with links to other regions.'

The entrepreneur as winner

To begin with the volumes were relatively small, with consequent financial and logistical problems. The ultimate success depended on the potential for up-scaling. MijnBoer consciously decided in favour of expanding, i.e. looking for more and other customers. The most important consideration behind the choice to expand was that of achieving a financially sound business. In this regard people were aware that a long-term vision (with financing plan) was required for the development of a robust organisation.

The robustness of an organisation is the establishment

of an organisation in such a way that it is able to respond quickly and flexibly to new, as yet unknown developments while still being able to offer existing products and services competitively.

MijnBoer proved adept at identifying and capitalising on opportunities that arose within the network of which it was part. Sometimes, however, there was a lack of awareness of the consequences of particular choices for the business model.

A long-term vision can act as a guideline for whether or not to take up certain opportunities. What is important is for that long-term vision to be consistently placed on the agenda and to exchange ideas on such questions as *'Where do you want to go?'* and *'What is your dream?'*. This provides clarity for people both inside and outside the organisation as to where the business is heading and what steps need to be taken to that end.

Financial security was an important factor in making choices. In the initial stage, MijnBoer was partly dependent on subsidies and mainstream business. In addition it is important to have long-term agreements with an investor or a bank, so that you are not dependent on arranging financial security when opportunities arise. As it is, certain choices may be based on the generation of revenues in the short term, rather than building on the long-term vision.

6.8 The present challenges

The new collaboration with Sligro/Smeding provides MijnBoer with access to knowledge concerning communication strategies. An overall strategy under the 'Honest and Delicious' strategy will allow MijnBoer to evolve into a strong brand.

Marqt sells fruit and vegetable products of MijnBoer on a sale and return basis. In the photo: Quirijn Bolle, managing director of Marqt (left) and Marco Duineveld, founder of MijnBoer B.V. (right) (Photo: Mugmedia, Wageningen)

The entrepreneurs of the Rondeel: Ruud Zanders (left) and Gerard Brandsen (Photo: Mugmedia, Wageningen)

7. RONDEEL

eggs as chickens like laying them

7.1 The challenge

Intensive livestock farming is the regular subject of heated debate in Dutch society. All sorts of values, attitudes and facts play a part in this debate. Many members of the public for example indicate that they regard animal welfare and sustainability as important. At the same time the price of food is often the decisive factor in consumers' choice, and in order to survive, farmers are forced to put profitability first. To date the choice of eggs was primarily that between animal-friendly and more expensive (organic and free range) or and animal-unfriendly and cheap (conventional, barn). Conventional chicken farming has a high impact on the environment as it requires a lot of energy and the manure generates a great deal of ammonia and nitrogen. Organic farms also fail to take any – or sufficient – account of manure production and energy consumption.

Cor van de Ven (owner of the *Venco Group*, a supplier of housing systems for the poultry sector) asked himself: *'How can I respond to the growing demand for both environmental-friendliness and animal-friendliness?'* Cor van de Ven had the vision of launching new housing systems in the market that would be able to continue satisfying all the public's wishes in the future. His aim was to meet the wishes of 'citizen, farmer and animal'. This meant that new quarters had to be devised that meet the requirements of the animal, are valued by the general public, keep animals healthy and minimise emissions. The challenge for Rondeel was to capture a place in the market for an egg that costs more to produce than conventional eggs, even though animal welfare is often a vague concept for both retailer and consumer.

Collaboration was sought to that end with various parties: poultry farmers, supplier of housing systems, retailers, NGOs, government authorities and research institutes. This gave rise to an approach from which we can learn how a commercial value was attached to environmental-friendliness and animal welfare in the market.

7.2 How did the innovation come about?

The basis for the innovation was laid in 2003 with the technical development of the housing system. The planning phase provided a clear target to aim at. The TransForum project (2007-2010) was primarily aimed at the actual development of a Rondeel laying hen housing system. Particular attention was paid to the organisation of the new production and distribution chain. The concept of the Rondeel housing system was then refined in talks with the Animal Protection Foundation.

An explanation of the technical design of the housing system is provided below, followed by how it proved possible to move from design to commercial realisation.

A new type of collaboration was required in order to turn the Rondeel into a commercial success. All sorts of parties had to be involved in order to acknowledge the added value of the housing system (in terms of animal welfare and the environment). Only then was it possible to realise added value.

Project partners
Animal Protection Foundation, poultry-farmer Gerard Brandsen, Rondeel B.V., Venco Group, Wageningen UR (Livestock Research and Business Administration) and TransForum.

TransForum project
2007-2010

The start:
the technical design of Rondeel

The technical concept of the Rondeel was developed as part of the Ministry of Agriculture, Nature and Food Quality project The Keeping of Hens (2003-2004). Wageningen University was involved in the research into the welfare of chickens.

The design takes full account of social issues and needs such as sustainability, animal welfare, openness and education, harmonising with the landscape, greater profitability for the entrepreneur and closer cooperation among the chain-players.

The sustainable housing system for laying hens meets the animal's needs, with specific areas designed to meet specific needs.

Entrepreneur Ruud Zanders
(Photo: Mugmedia, Wageningen)

- *Night quarters:* Once it is dark the chickens go to the night quarters. This is where laying hens sleep, eat, drink and lay their eggs. These quarters largely make use of existing techniques for keeping laying poultry.

- *Daytime quarters:* This is where the laying hens scratch, rest and play. Here they are able to scratch, scrape and take dust-baths to their hearts' content. Nature has been brought inside. The side-walls separating the night and daytime quarters can be rolled up. In this way a single climate is created, so that more hens make use of the scratching space in the daytime quarters. The hens are also able to go outside from the daytime quarters to scratch and scrape at the wooded fringe.

- *Wooded fringe:* The outer perimeter, with natural grass and vegetation. The hens are able to reach the woodland edge from the daytime quarters. This part can be easily closed off if an outdoor ban is introduced.

- *Central core:* The central core of the housing system consists of three storeys. The ground floor is the working space with the egg-packaging line. It is important for the poultry farmer to be able to keep an eye on the laying nest boxes from the central area. The space provides a pleasant working environment.

  The eggs are transported from the laying nest boxes by belt to the packaging machine. The eggs are packed straight away into the cartons. The first floor is the visiting area, which is open every day. It also provides meeting facilities for businesses. The second floor houses the heat-exchangers.

- *Visitors tunnel:* A visitor area provides a view of all parts of the farm. The Rondeel is open to visitors every day (except Sundays). Visits are made by all sorts of groups, such as schoolchildren interested in all aspects of the environment and sustainability, and fellow entrepreneurs, etc, who are highly interested in the new housing system. The visitors tunnel provides an eye-level view of the chickens as they scratch and scrape.

New types of cooperation make the innovation possible

The Animal Protection Foundation was actively involved right from the preparation of the technical concept in 2003-2004. This made it attractive for Cor van de Ven to work this concept up into a commercial product.

As part of the TransForum project 'The quest for the Golden Egg', the challenge was taken up in 2006 of translating the technical Rondeel concept into a commercial product. As noted above, marketing a sustainable egg calls for not just technological but also organisational innovation.

This requires interaction between practical know-how and scientific and technological knowledge. Combining various types of knowledge was important for the solution of complex problems in the field of sustainability. In this project the researchers were important for contributing scientific knowledge, the supplier of housing systems for providing the technical knowledge, the NGOs for generating the public and political support and the Rondeel entrepreneurs and the poultry farmer Gerard Brandsen for providing the knowledge concerning the commercial feasibility of the concept.

At the same time the Rondeel worked hard on establishing support for the new egg concept. Talks were therefore held with Ministry of Agriculture, Nature and Food Quality, suppliers and retailers. At local level the Municipality of Barneveld provided support by dealing flexibly with the planning procedures. Together with the Farmers Organisation ZLTO and the TS Consult consultancy, steps were taken to penetrate the national and international markets.

Hallmarks as an independent value judgement

1. Better Life hallmark

Thanks to close cooperation with the Animal Protection Foundation the new housing system for laying hens meets the strictest animal welfare standards. The Animal Protection Foundation awarded the Rondeel egg the maximum number of three stars under its Beter Leven ('Better Life') hallmark, when the first egg had still to be

The Rondeel mission

In consultation with the government, NGOs and the business community producing and selling eggs responsibly under the slogan: 'Many wishes and needs with one total solution!'

laid. This was a unique position, which had a knock-on effect among retailers, who were willing to buy the more sustainably produced egg provided there was sufficient quality assurance.

The three stars were awarded after the Foundation was also scientifically convinced of the added value. The award was made on condition that the added value would also apply in practice, which proved to be the case.

2. Milieukeur foundation

Partly on account of the application of various technical innovations (including the first manure carousel, heat-exchangers and natural ventilation) the Rondeel is the first housing system for laying hen to have been certified by the environmental hallmark foundation Milieukeur. Milieukeur lays down criteria for animal welfare, as well as for reducing environmental pollution in poultry farming. The environmental requirements are based around reductions in the emissions of nitrogen and phosphate, a greater number of low-emission animal spaces, and measures with regard to feed and manure. The animal welfare requirements concern a maximum level of loss from death, frequent visits by veterinary surgeons and providing the birds with extra living space.

Shared responsibility in the chain

In 2008 and 2009 a number of working sessions were organised as part of the TransForum project to look at some important issues: new roles for chain partners, marketing, dealing with the media, and the progress being made. Shared learning was key. The biggest gain was the development of a more open consortium, where the group process was an important factor. A shared goal and shared responsibility were needed in order to achieve the final aim. The individual interests were incorporated within this.

Rondeel involved a shift from *'what are you doing so that I can also do my thing'* to *'what am I doing so that we can all reach the finishing line together as best as possible?'* The openness meant that the real questions and also thorny issues were raised. For example, *'Who takes which risks (how transparent dare we be)?'*, *'What is expected of the pioneering role of Rondeel B.V.?'* and *'Who is co-owner of the concept?'* The poultry-farmers opted to participate fully in the development of the concept.

The Rondeel B.V. expressly put itself forward as the driver. Poultry-farmers then made it clear that they wished to take responsibility for the sorting, packaging and information functions. Detailed attention was paid to setting up a new, short chain, with new roles and responsibilities for the players.

The Venco Group regarded the first Rondeel housing system primarily as a demonstration system. Since it did not prove possible to attract finance in the capital market for the first Rondeel system, the Venco Group decided to finance it itself. In doing so it also took the full responsibility for the risks. Rondeel B.V. is the organisation that is responsible for the name recognition of Rondeel and the marketing of the eggs. The bulk of the egg production will be marketed nationally. The remainder of the eggs will be sold primarily by means of door-to-door sales, for which the poultry-farmer will be responsible.
The chain is now highly transparent for the poultry-farmer,

who is familiar with all the selling prices. The poultry-farmer obtains a percentage of the extra return, so that it is in his interest to share in the marketing and promotion of Rondeel. The poultry-farmer does not just keep laying hens but also lets out the meeting room above the Rondeel itself, and shows visitors via the visitor tunnel the chickens live: sustainability also means full transparency. This has proved a great success, with many and also enthusiastic visitors. Poultry-farmers become the co-owner of the Rondeel company, and hence also an ambassador for and seller of this concept. Where previously the egg-chain had traditionally been organised in one particular way, a new business model has arisen based on shared ownership. Rondeel eggs are packaged on-site and transported directly to the supermarket. An egg-packaging station no longer forms part of the chain but is integrated into the poultry farm. This makes the logistical process simpler and cheaper. It has been made possible for the poultry-farmer to buy the housing system in a few years' time, once it has become clear that the Rondeel concept is successful. The Venco Group would like by that time to concentrate on its core business again, namely selling housing systems for the poultry sector. It does not have to remain responsible for the marketing of eggs in the longer term.

'The wooded fringe', part of the daytime quarters

(Photo: Mugmedia, Wageningen)

2003	Crisis situation before outbreak of avian influenza
	Minister of Agriculture Cees Veerman issues instructions for the redesign of a laying hen housing system
2004	Project 'The Keeping of Hens' based on a multi-stakeholder approach
	Design of Rondeel by Wageningen University-Livestock Research in collaboration with the Animal Protection Foundation
2006	Start of TransForum project with Vencomatic, Animal Protection Foundation and Wageningen University
2007	First contacts with poultry-farmers
2008	Establishment of Rondeel B.V. and appointment of managing director (Ruud Zanders). This made it clear to all concerned that Venco would continue seriously
	Discussions with the Animal Protection Foundation to refine the concept
	Working sessions on the organisation of the chain and new partnership models
	Active interaction with government authorities for building and environmental permits
2009	Contract with poultry-farmer Gerard Brandsen in Barneveld
	Collaboration entered into with Farmers Organisation ZLTO for the marketing of eggs
	Lobby for risk capital (2008-2009): No support forthcoming from banks or ZLTO
	Application for Ministry of Agriculture financial guarantee scheme: not granted
	Lock-in: Venco Group decides to finance the construction of the first (demonstration) Rondeel itself
	Organisation of new chain: poultry-farmer as co-owner of the concept
	Permits granted by the Municipality of Barneveld
	Construction commences on the first Rondeel
	Three stars of the Better Life hallmark awarded by Animal Protection Foundation
	Ministry of Agriculture financial guarantee scheme awarded for the construction of second, third and fourth Rondeel
2010	Opening of first Rondeel in Barneveld in April
	First eggs on the market in June
	Contract with Ahold
	First egg with Milieukeur hallmark
2011	Opening of the second Rondeel in Wintelre in February

Rondeel

Rondeel is genuinely distinctive as it is an integrated, sustainable chicken shed. Every detail has been taken into account. The result has been a positive shift in favour of animal-friendly eggs in Dutch supermarkets.

➕ The advantages

Poultry-farmer

Rondeel BV amounts to a unique relationship between supplier of housing systems and poultry farmer. The chickens occupy a pleasant and spacious shed.

Retail

A contract with the supermarket chain Albert Heijn guarantees the sale of Rondeel eggs.

Consumer

On offer is an egg that has been produced sustainably and under good animal welfare conditions.

➕ Blends in with the landscape

➕ Animal welfare & animal health

➕ Minimal emissions & energy consumption

➕ Agreements about new division of the margin

➕ Farmer able to invest in sustainability

➕ 3 stars from Animal Protection Foundation & Milieukeur eco-label

➕ Room for natural animal behavior

➕ Consumers are able to visit the Rondeel

➕ Environmentally-friendly packaging

Layout

Central Core
(air extraction and ventilation)

Night quarters
(sleeping, feeding, drinking and laying eggs)

Daytime quarters
(scratching, resting, playing and hiding)

Dust bath
(scratching and scraping)

Wooded fringe
(hiding and scratching)

5m   75m   5m

	Shed size (building)	Free range area	Sales breakdown of Dutch eggs
BARN	1,667 m²	0 m²	84%
FREE RANGE	1,667 m²	75,000 m²	3,3%
ORGANIC	2,500 m²	120,000 m²	1,9%
RONDEEL	4,400 m²	5,700 m² (incl. wooded fringe)	0,08%

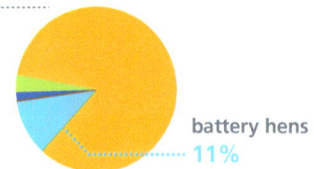

battery hens 11%

140

7.3 Key figures

- In the Netherlands per capita consumption is around 180 eggs a year (a little over three a week), including the eggs used in processed products.
- There are over 45 million laying hens in the Netherlands, laying nearly 10 billion eggs a year.
- A Rondeel houses 30,000 laying hens (5 daytime/night quarters each with 6,000 chickens).
- The average diameter of the Rondeel is 75 metres, and including the wooded fringe 85 metres.
- The Rondeel covers an area of 4,400 m^2, and including the wooded fringe edge 5,700 m^2.
- A Rondeel can therefore be built on a plot of less than one hectare.

A comparison for 30,000 laying hens

- Traditional barn hens, one floor: 1,667 m^2 of space (building area).
- Free-range facilities: 1,667 m^2 of space (building area) + 7.5 ha free-range facilities.
- Organic: 2,500 m^2 of space (building area) + 12 ha free-range facilities.
- Of the total number of eggs in 2007 84% were barn, 11% battery, 3.3 % free-range and 1.9% organic.
- 84% of these eggs were sold in supermarkets.
- Of the supermarkets Albert Heijn has a market share of 31%.
- For a Rondeel the anticipated weekly production is 150,000 eggs.
- A Rondeel system accounts for 0.08% of Dutch egg sales.
- If all the eggs in the Netherlands were to come from Rondeel housing systems, 1,200 of these buildings would be required.

7.4 The added value of the Rondeel

The advantages of the Rondeel

The Rondeel is a genuinely distinctive housing system since it approaches sustainability on an integral basis: All aspects are taken into consideration, including packaging and manure production. Various sustainability features have been improved as a coordinated whole.

1. The farm is attractive to work in and is highly competitive.
2. The farm is open and accessible. Transparency concerning the way in which chickens are kept is an important element of the concept.
   From the visitor tunnel members of the public are able to view the chickens at eye-level. A path around the perimeter allows visitors to see how the chickens take a dust bath in the wooded fringe.
3. The Rondeel is a housing system for laying hens that blends into the landscape.
4. The farm meets the animal's needs in terms of both animal health and welfare. In the case of an outdoor ban the system can be sealed off.
5. The farm minimises emissions.
6. Rondeel housing systems use less energy.
7. The packaging used for the eggs is environmentally-friendly.

The sustainability performances

People

The farm is open and accessible. It combines professional poultry-farming with educational and recreational functions. Everything is aimed at a direct relationship with the consumer. Animals become visible (again) to the public, both in the landscape and on the farm. The farm has a licence to exist and is not a source of nuisance due to odour, noise or excessive transport movements.

The Rondeel is a laying housing system that blends into the landscape. Harmonisation with the landscape was a major issue in the design of the concept. A Rondeel rises from the landscape like a hill. It is surrounded by hedge banks and there is a planted wooded fringe.

Animal welfare

With regard to animal welfare, consideration was given to the aspects of stocking density and the dust bath area. The stocking density amounts to 6.7 hens per m^2. By way of comparison, an organic housing system has an occupancy of 6 hens per m^2 (excluding free-range facilities). At the present time some of the Rondeel hens receive beak treatment in the hatchery, while others do not.

The results after three months are positive: there is little pecking. The intention is to avoid the need for trimming in all further Rondeel systems. At the present time the effective breeding of chickens without beak treatment is an obstacle. In order to resolve this the plan is for the fourth Rondeel to be designed in such a way that it can be used to breed chickens that are suitable for the existing Rondeel system. The hens are given normal feed, with an extra dose of soya. In order to keep the laying hens occupied, the poultry-farmer scatters grain in the grass. The comparison of animal welfare has been based on the FOWEL study by De Mol et al. (2004), which compared differing systems in terms of 25 animal welfare attributes. Important factors in poultry-farming include space per chicken, availability of eating facilities, water, perches and laying nests, the ability to take dust baths, while there are also negative factors, such as beak-trimming and exposure to predators. At 9.6, the Rondeel, as designed in the Keeping of Hens project, was awarded the highest score, followed by organic with 7.8, free-range with 6.8 and barn 6.3. For the Animal Protection Foundation the outcome of this (independent) study was particularly important in its decision to award three stars under its Better Life hallmark.

Animal health

The Rondeel is a laying hen housing system with a wooded fringe where the chickens are free to move about. The wooded fringe can be closed off if an outdoor ban is imposed. This helps reduce the risk of the transmission of diseases (e.g. zoonoses such as avian influenza) from wild birds to human beings.

Planet

The farm minimises emissions

The ammonia emissions are in the bottom-most category of the Waste Incineration Directive (WID). Little ammonia is released in the Rondeel system as use is made of natural ventilation and the manure is removed and dried twice a week. The rapid drying of the manure combined with post-drying achieves an ammonia reduction of some 50%. The emissions of odours and particulates are consistent with the WID guidelines. The air used in post-drying the manure reduces particulates by around 50%. The natural ventilation means that there is no forced air flow, with a consequent reduction in particulates.

Energy

Less energy is used since the entire system is naturally ventilated. The Rondeel is well insulated; the pop-hole doors also take the form of insulated and integrated roller shutters (instead of uninsulated pop-hole hatches in traditional housing systems). An optimal group climate is obtained with the use of heat-exchangers.

Manure

Each chicken produces 15-20 kg of manure in the laying period. By making use of a manure carousel with an integrated post-drying system, the manure in the Rondeel is much drier than normal (with a dry substance content of around 80%), allowing it to undergo final processing straight away without the need for it to be treated as a waste substance. This of course also cuts down road transport costs, as less 'water' is transported. Talks being held with various parties who would like to market Rondeel manure are now at an advanced stage.

The packaging of eggs is environmentally-friendly

The eggs can be directly sorted and packaged on the farm

for the consumer. The eggs are packed into a round, naturally biodegradable coconut-fibre carton containing seven eggs and bearing the Albert Heijn supermarket sustainability label *puur&eerlijk* (pure&honest) and Rondeel logos. The Rondeel lends itself ideally to sorting and packing 'on the farm', thereby eliminating unnecessary transport.

Profit

The farm is attractive to work in and is highly competitive commercially. A poultry-farmer is at once a labourer, entrepreneur and animal-keeper. The Rondeel provide space for professional expertise and pleasure in one's work. The poultry-farmer himself scatters grain in the henhouse every day and so also has genuine contact with the birds. The continuity of the business, production certainty and a decent income are all important. Caring for the animals is a key part of his work. Labour-saving is achieved by means of modern packaging machines and an improved view of what is happening in the housing system. The poultry-farmer also benefits substantially from the higher price paid for Rondeel eggs.

The Rondeel investment costs are substantially higher than those for other laying hen housing systems. The investment per bird is over twice as high in comparison with an aviary and free-range and at least 30% higher than organic housing systems. Eggs are by way of bulk products, and achieving a higher price was not easy. To date no one has been prepared actually to invest in a new design. There's nothing to say that you will ever be able to recoup your investments… Here we have an entrepreneur with courage and a vision for the future. The Venco Group is a very large supplier of housing systems, which also operates internationally. The Rondeel would like to continue developing internationally. Talks are being held in Germany and Belgium with NGOs and retailers.

Construction of the first Rondeel in Barneveld (Photo: Bart Jansen)

SWOT analysis of the sustainability performances

Strengths

- In comparison with organic and free-range eggs, a positive effect on animal health.
- In comparison with barn eggs, a positive effect on animal welfare.
- A positive contribution towards community involvement through greater transparency and accessibility.
- A good financial outcome for the poultry-farmer.

Weaknesses

- Land-use per egg is higher as more feed is consumed for the same number of eggs.
- In comparison with barn eggs, more food is needed for the same number of eggs.
- Realising a new system requires higher investment costs (on account of new techniques and marketing).

Opportunities

- The introduction of requirements for the supply chain, for example insisting that feed manufacturers use sustainably produced based materials (e.g. with an EKO or Milieukeur hallmark, or sustainable soya).
- The Rondeel is seen as having positive potential on account of the project's scalability.

Threats/risks

- The stability of the price of the egg.

7.5 The value creation model

Notes on the value creation model

The Rondeel has been able to create value by developing the concept in collaboration with scientists, the Animal Protection Foundation and the Milieukeur foundation. Public support has been generated by organising a transparent chain and putting together a network organisation. Recognition of this support has been turned to account by the award of

animal welfare and environmental hallmarks (Better Life and Milieukeur). This independent quality assurance provides the Rondeel egg with a unique selling point for the supermarket chain Albert Heijn, and enables Rondeel to become a preferred supplier. Housing system supplier, poultry-farmer and retailer jointly take responsibility for the marketing and sale of the eggs.

The creation of values may be shown as follows in the model (see alongside).

7.6 From plan to investment

Financing of the first Rondeel

The construction of the first Rondeel cost several million euros. Financial commitments by the Ministry of Agriculture, Nature and Food Quality and banks were a long time in coming. The Venco Group therefore decided to undertake the investment in the first Rondeel itself, in the knowledge that a demonstration Rondeel was required for the concept to be taken further. A financial guarantee provided by the Ministry of Agriculture, Nature and Food Quality for the second Rondeel helped persuade other investors to take the plunge.

Earning money from sustainable products

Animal-friendly production and the Milieukeur hallmark generate market value. In the field of animal welfare there was much room for improvement in the laying hens sector. Until 2010 there were just three categories of eggs: organic, free-range and barn. The categories were based on the number of chickens per square metre. The Rondeel housing system differs from the other categories in that it is an integrated, sustainable coop.

The fact that the egg is also genuinely sustainable is evident from the award of hallmarks. As noted the Rondeel egg has been awarded three stars by the Animal Protection Foundation under its animal welfare hallmark Better Life and is the first egg to have been awarded a Milieukeur label. The supermarket chain Albert Heijn therefore regards the egg as having a unique selling point, and has added it to its

Value creation model
Rondeel

Competencies
- Market orientation
- Entrepreneurship

Co-creation with Animal Protection Foundation

Communication with local community and stakeholders

Investments
- Rondeel housing system
- Visitors' facilities
- Certification

Transparent entrepreneurship, sharing risks in the chain

Housing system optimisation

Investment in own packaging lime

Unique Selling Points (USP)
- Egg with respect for animal welfare
- Environment-friendly egg
- Housing system open to the consumer

Branding through use of hallmarks

Results
- Albert Heijn supermarket preferred supplier
- First egg to have Milieukeur environmental hallmark
- Animal Protection Foundation 'Better Life' hallmark with three stars
- Rondeel egg with puur&eerlijk sustainability label

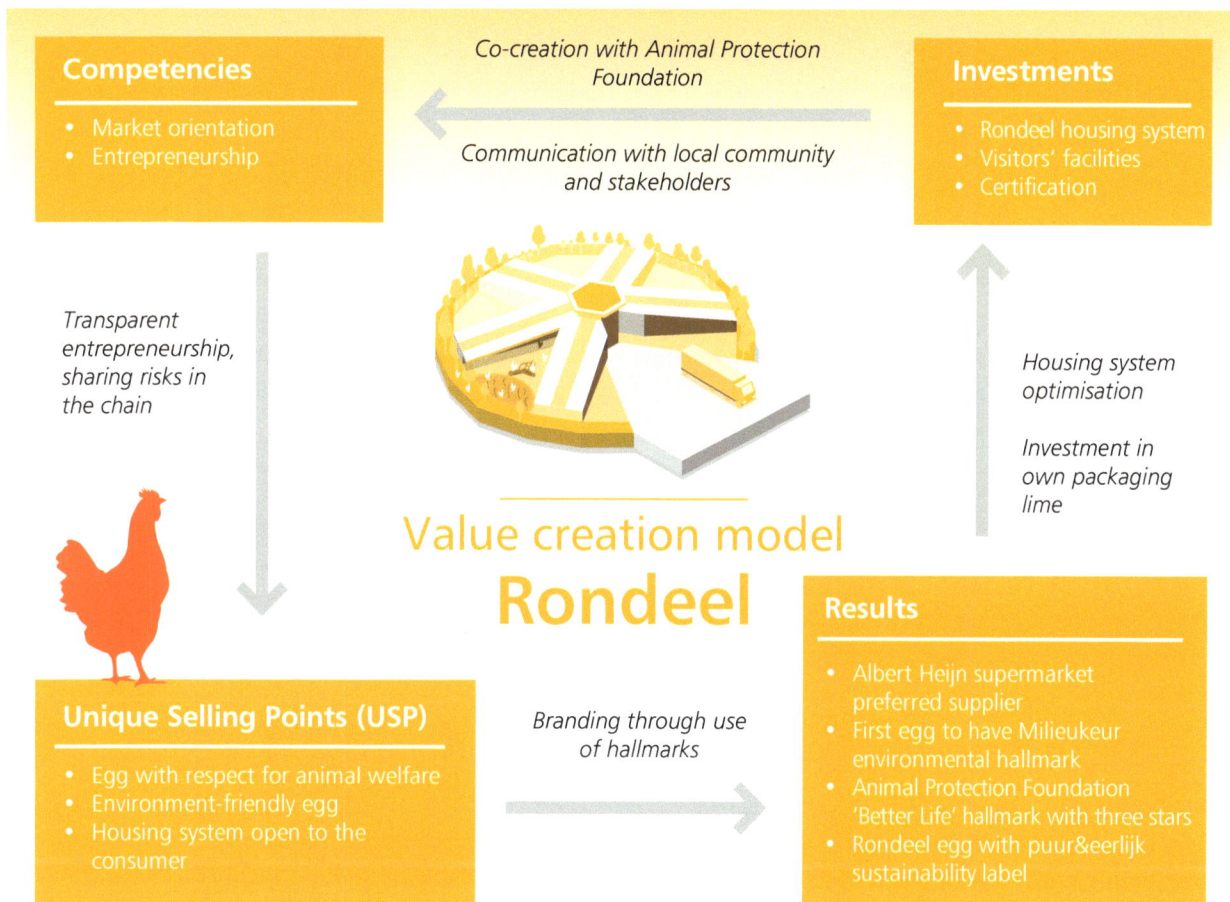

sustainability label puur&eerlijk. Contracts have now been signed with Albert Heijn. Rondeel eggs sell at 1.89 euros for a carton of seven eggs (reference date 2010), which is below the price of organic eggs. Albert Heijn accounts for the bulk of the sales; the remainder of the eggs will be sold by means of door-to-door sales.

Transparency within the chain

A breakthrough has been achieved in the retail contract. For the first time, the selling price of eggs is being linked to the price of chicken feed, the biggest variable cost item in producing an egg. By establishing a link between the selling price and the feed price, the poultry-farmer is paid a fair price, so that investments can be recouped. This would appear to be an important step towards a genuine partnership with the retail industry.

7.7 The lessons for the entrepreneur

The entrepreneur as project developer

The planning phase formed part of the project 'The Keeping of Hens', where a technical design was worked out. This phase provided a clear target to aim at. The lesson is that a lot of time, energy and courage are needed to convert a technical concept into an innovation capable of making its way in the market. The investment phase formed part of the TransForum project and was aimed primarily at the establishment of a network organisation and the actual development of a Rondeel system. The entrepreneurs concentrated particularly at that stage on the organisation of the new chain. The operating phase involved the construction of the first Rondeel and the sale of the eggs by supermarket chain Albert Heijn.

The entrepreneur as coach

Reflection helps one stand back, so that the concept can be placed in a broader perspective. It's useful to learn from other sectors. In order to respond effectively to social developments it is necessary to take a broad view at all times and to step beyond the confines of one's own environment. New insights were generated by sparring with one another and with other innovative projects. This was done by the entrepreneur in this project by organising a working session several times a year on an important topic and discussing this topic with external experts.

The entrepreneur as strategist

A highly important success factor for this innovation was the vision on the part of the initiator, the Venco Group, that a new housing system needed to be sustainable and animal-friendly, and also that public support needed to be generated in order to recoup people and planet investments. Efforts were therefore made to commercialise quality and sustainability.

The entrepreneur as games-leader

The organisation of a new, short and transparent chain resulted in a new type of business model based around joint ownership and responsibility for the marketing of eggs. This called for trust, and all the participants were involved in the process.

The entrepreneur as spider in the web

The Venco Group was aware that a new approach was required in order to turn animal welfare and environmental performance into a commercial asset. Cooperation with the Animal Protection Foundation was vital in this regard. A broad network (governments, research institutes and the farmers organisation ZLTO) was also required to build up public support for the concept. Arriving at a new type of agroproduction involves not thinking immediately in terms of a solution but reflecting on what is going on in society. An open mind is required. An innovative concept of this kind – an egg based on emotion and welfare – is not something that can be developed by one player alone, let alone launched in the market. It calls for new insights, new knowledge and new incentives, for example to get away from old ways of thinking (through the co-creation of knowledge).
A multi-stakeholder approach means that the full range of interests can be weighed. By building up the right network it becomes possible to generate public and political support. This means keeping in step with the market and the general public.
The setting up of the Rondeel involved highly active networking with all the stakeholders concerned. Strategic cooperation with non-governmental organisations has made it possible for Rondeel to charge a higher price for its eggs and so pay off the investments in sustainability. The lesson: different kinds of non-governmental organisations need to be involved at an early stage.

The entrepreneur as winner

A new intrinsic product value calls for a different positioning in the market. Rondeel meant working on an egg based around welfare and emotion. This calls for non-conventional backers or ambassadors. It also calls for additional emphasis on value creation.
Strategic cooperation with non-governmental organisations has made it possible for Rondeel to charge a higher price for its eggs and so pay off the investments in sustainability. The new type of business model is characterised by collective ownership, transparency and shared responsibility for the marketing of eggs. Making the shift from old (chain-based) ways of thinking to a commitment to a total reorganisation of the chain takes time but is imperative.

The division of roles in the project must be clear. Rondeel's coordinating role provides clarity and belief for the parties concerned and also conveys strength to the outside world. From the outset consideration was given to the strategy that would enable the investments to be recouped. The first Rondeel came about as a result of investments made by the Venco Group. A financial guarantee provided by the Ministry of Agriculture, Nature and Food Quality persuaded other investors to take the plunge.

In the meantime Rondeel is continuing to build. The second Rondeel system opened in February 2011. Plans for a third Rondeel are on the drawing board. Five Rondeel systems are scheduled to be completed in 2013. Rondeel is currently also in talks with German and Belgian partners. Once again the strategy here is to do so with both retailers and NGOs.

Ruud Zanders
(managing director of Rondeel):

'From the point at which the decision to build was taken, - so "we'll just have to go ahead without the Ministry of Agriculture", and "in that case we'll have to do without a sales contract in advance " - everything went into higher gear.'

7.8 The present challenges

The first Rondeel housing system for laying hens was ceremonially opened in Barneveld on 8 April 2010 by the managing director of Albert Heijn, Sander van der Laan: a milestone in the process from design to realisation of a sustainable housing system.

Core promise for the consumer:
'Thanks to Rondeel I can always honestly enjoy the most animal-friendly eggs'

The first eggs went on the shelf at Albert Heijn in June 2010 under the sustainability label puur&eerlijk. The offtake of eggs is guaranteed under a contract with supermarket chain Albert Heijn. The public debate can affect the sales of various types of eggs. The will exists at Albert Heijn to turn the Rondeel egg into a success. Ultimately it is the consumer who will determine the outcome.

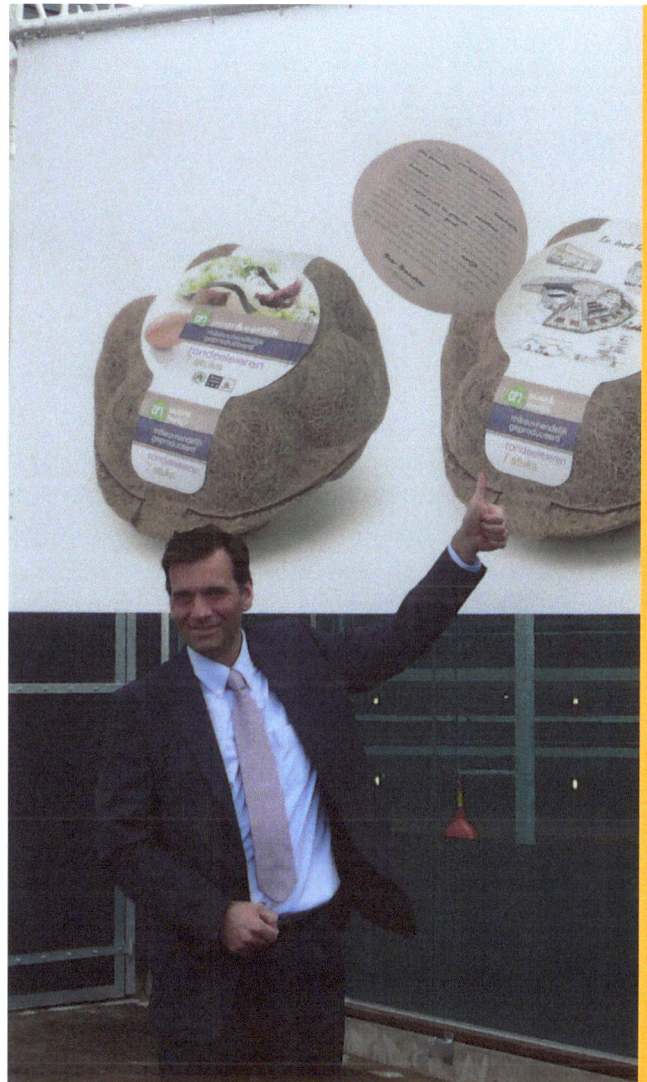

Albert Heijn includes the Rondeel eggs in its range In the photo Sander van der Laan, managing director of Albert Heijn

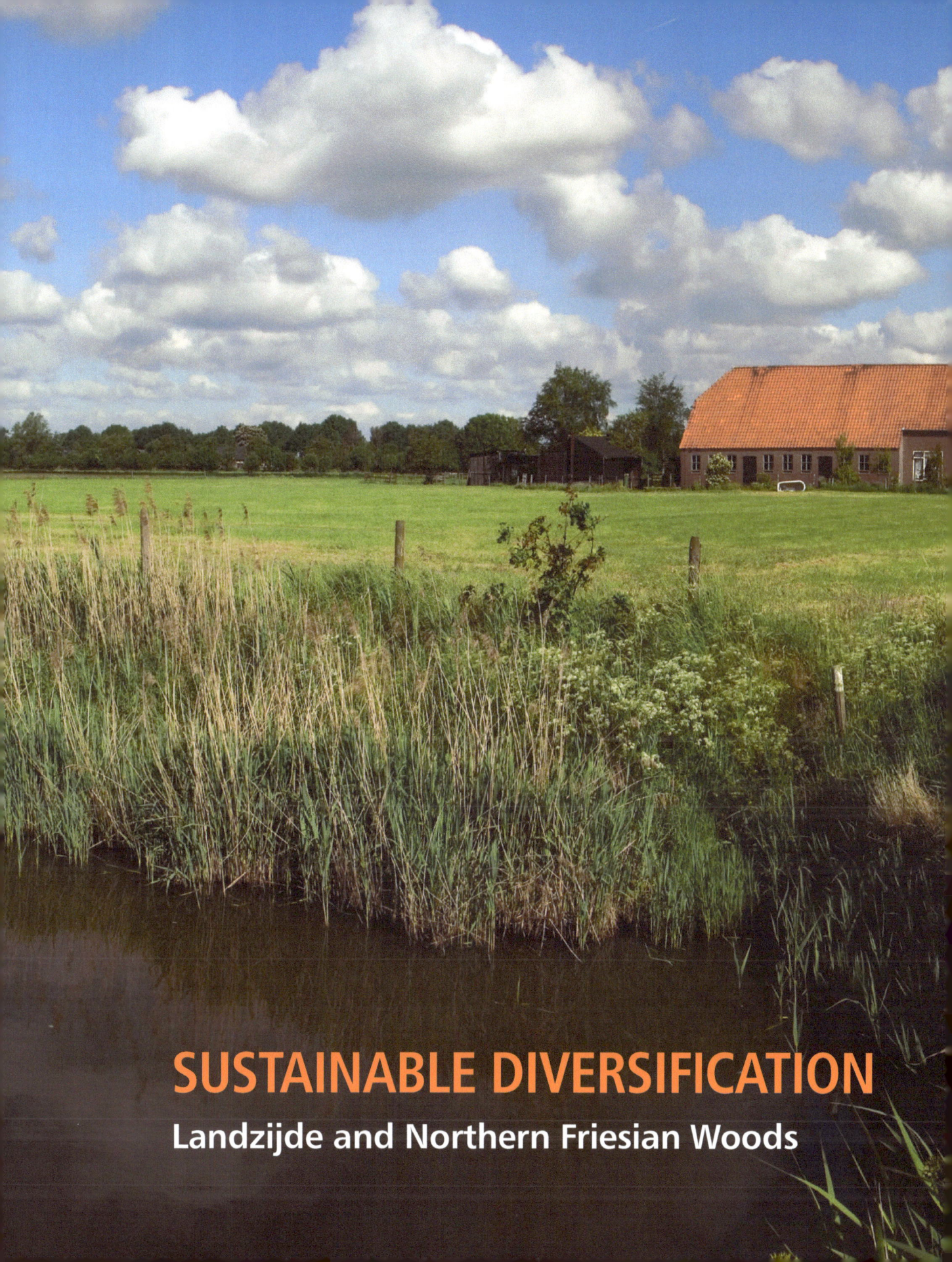

SUSTAINABLE DIVERSIFICATION

Landzijde and Northern Friesian Woods

The care farm gives people a fully-fledged place in society again

8. LANDZIJDE

farmers help with care

8.1 The challenge

'How do you open up the new 'care market' for farmers in Waterland?' That was the challenge for *Jaap Hoek Spaans*, a farmer and former teacher, when he founded Landzijde in 2000. The growing urbanisation imposed limitations on the farmers around Amsterdam, while also offering new opportunities. Waterland, an area to the north of Amsterdam, is increasingly regarded as an 'overflow' area for the residents of Amsterdam. They attach particular value to the old Dutch cultural landscape of peat meadows and abundant water. The ancient Beemster and Schermer polders have a place on the World Heritage List and were designated as National Landscape in 2004, meaning that the landscape must be preserved. As it was, the peat land areas and the abundance of the water didn't make life easy for the farmers. Now, they also faced the restrictions on landscape use. But the proximity to Amsterdam also provides an ideal 'market' for farmers wanting to diversify.

Landzijde Foundation wanted to develop a chain linking the demand for care in the city and the availability of peace, space and greenery in rural areas. This called for the professionalisation of care farming.

The Landzijde vision
'The clients are central and professionalism is the point of departure.'

organisations, care insurers and research institutes were also given a place in the network. The necessary start-up subsidies were also obtained at this stage. Landzijde and the research institutes worked on the development of their network by means of bilateral meetings, presentations and lectures, the hosting of excursions and participation in all sorts of regional consultations. They also worked on identifying the opportunities for and threats to care farming in the region.

Market-oriented instead of supply-led

In the development of the network and the chain, Landzijde consistently took the client (i.e. people with a disability) as the starting point. Landzijde consciously approached this not from the supply side (the care farmer) but from the demand side: the client and the care institutes.

8.2 How did the innovation come about?

The Landzijde mission
'Supporting people with a disability and making sure that they find a place with care farmers, so as to give them a fully-fledged place in society'.

The start-up: focus on network formation and chain development

In the interests of developing an effective product-market chain, Landzijde established connections with care institutes in Amsterdam, while at the same time forging a local network of farmers in the region. Municipalities, the province and the state, the political system, national and regional care

Project partners

Centre for Agriculture and the Environment, Municipality of Amsterdam, Koninklijke Nederlandse Heidemaatschappij, National Community Work Centre, AGORA Foundation for Special Education, Landzijde Foundation, Trimbos Institute, VU University Amsterdam (Athena Institute), Wageningen UR (PRI) and TransForum.

TransForum project
2005-2009

2000	Landzijde established by Jaap Hoek Spaans
2003	General Exceptional Medical Expenses Act (AWBZ) recognition of Landzijde as a care institute
2005	Agreement with National Agriculture and Care Support Centre: Landzijde pilot project
2006	Start of TransForum Green Care project
	Landzijde sphere of activity designated as a European pilot
	Connections with Streetcorner and welfare work established through workshops and discussions
	Daytime activity projects for homeless drug addicts from Amsterdam on two farms
2007	Two schools for special education spend time at a care farm and children are observed
	Reflection by Athena Institute on the project results in a new approach: combination of network-oriented and project-oriented working
	Commencement of meetings with municipality and care institutes
	Commencement of TransForum scientific project to demonstrate therapeutic effects
	Minister of Agriculture Verburg awards first diplomas for senior secondary vocational education course in care farming management (Groenhorst College in collaboration with Landzijde)
2008	Joint scientific article with professor of Mental Health Care and consultation with the Health Council give Landzijde access to the 'established medical order'
	Municipality of Amsterdam starts 'Amsterdam Experimental Garden', with Landzijde participation
2009	National Agriculture and Care Platform established, Landzijde focuses on national professionalisation of care farming
	Landzijde moves into self-contained new premises in Purmerend: the former Burgerweeshuis (Civic Orphanage)
	Start of TransForum scientific project: Business models in care farming, based on the need for the further professionalisation of care farming.
2010	Landzijde sets up its own monitoring and effect measurement system

The organisation of a chain and building up and maintaining networks did not in themselves create any value. For this a business model had to be set up that created value and was self-sustaining.

The Landzijde Foundation business model may be characterised as that of a broker between care demanders and care farmers. In the course of the project the organisation was developed further: Landzijde moved into new premises, set up an administration system, recruited care coordinators and concentrated on the Province of North-Holland as a whole with the appointment of regional coordinators. The organisation not only linked up patient care and farm upkeep but also took steps to promote the quality of new knowledge development (by for example courses and discussion evenings), among care farmers themselves (intervision), between care institutes and care farmers and between the participants and the knowledge institutes.

Creating value for all stakeholders

With the aid of researchers Landzijde formulated what the **clients** considered important: at issue were meaningful daytime activities in order to re-establish a fully-fledged place in society. This therefore concerns a place on a 'real' operating farm with a 'real' farmer and his wife, instead of a place within an institution with carers and therapists. The requirement for the care farmers is therefore that the farmer remains a farmer, is authentic and remains the owner of the farm.

The **care institutes** have requirements with regard to the standard of care, the organisation of transport, safety on the farm and the quality of the support provided by the farmers. These requirements are satisfied by means of care protocols drawn up for each target group, quality and safety requirements on the farm, insurance and intake, support and guidance, training and assessment of the care farmers, all this leading to a certification system. Landzijde accordingly became a quality guarantee or 'strong brand'. For the care institutes and the care farmers it turned out to be important continually to measure whether the clients on the farm were satisfied with the care on offer. For this purpose Landzijde developed a monitoring system. These quality measurements enable account to be rendered to the care institutes and financiers, and enable to improve the standard of care on the farms.

The participating **care farmers** want an increase in income commensurate with the additional work and clients that are suitable for the farm – all this without undue administrative hassle. In brief, a well-oiled organisation relieving them of as much additional organisational work as possible.

The **financiers** are seeking the efficient and effective utilisation of their budgets. Apart from that they are also interested in the satisfaction of their clients. They therefore require accountability and monitoring systems, satisfaction measurements and the certification of the care farmers.

A social role for the farm in the community led to increased acceptance

Landzijde concentrated not just on its core business but also profiled itself as an organisation that wanted to strengthen the links between city and countryside. To this end it joined in with the municipality's Amsterdam Experimental Garden initiative. Landzijde farmers receive school classes on the farm, support other school projects and are present at regional and farmers' markets in the city.

Apart from increasing their network as a result, this contributed towards the name recognition and public acceptance of Landzijde as an organisation of value to the city.

Critical reflection within the project development in order to retain focus

Within the TransForum project, a member of the Athena Institute of VU University Amsterdam was given the role of 'reflector'. She regularly held a mirror up to the project and kept a record in meetings of the questions that needed to be answered. The director of Landzijde indicated that this made him aware of strengths and weaknesses, and answers were formulated during the reflection to the questions that had arisen. The reflection prevented tunnel vision within the project.

8.3 Key figures

- In 2009, 102 farmers were affiliated to Landzijde and care was provided for 421 clients (on average 101 half-days per client).
- Farmers' incomes were between € 3,000 and € 180,000, with an average of € 70,000 a year.
- Landzijde's turnover from care activities amounted in 2006 to 1.1 million, in 2007 to 1.9 million, and in 2008 to 4 million.
- The average cost per client for Landzijde amounts to € 4,657 a year.
- In 2008 the Netherlands had 800 care farmers, approximately 1% of the total number of farmers.
- At present around 10% of this potential market share in the Netherlands is realised by the care farming sector.

8.4 The added value of Landzijde

The advantages of Landzijde

1. Landzijde organises quality assurance and monitoring protocols for clients and care institutes in respect of the care provided on the farm.
2. The organisation acts as a placement agency for care farmers.
3. It avoids administrative hassle for the farmers.
4. Acquisition, recruitment and placement are undertaken centrally.
5. Landzijde offers farmers an ongoing range of courses and training.
6. It is a large organisation and hence an effective and influential discussion partner for policymakers and organisations.

1. Quality assurance for the care offered on the farm

Landzijde conducts intake interviews with prospective farmers. These are used to discuss the required competencies,

Landzijde

Landzijde supports people with a disability and helps place them with a care-farmer. How does this work?

The advantages

Care insurers

Financing by care insurers.

financial flow

Care institute

Customer for services provided by Landzijde and care farmers.

Landzijde

Central player acting as intermediaries between care institutes and care farmers.

Care farmers

Provide daytime activities and rehabilitation.

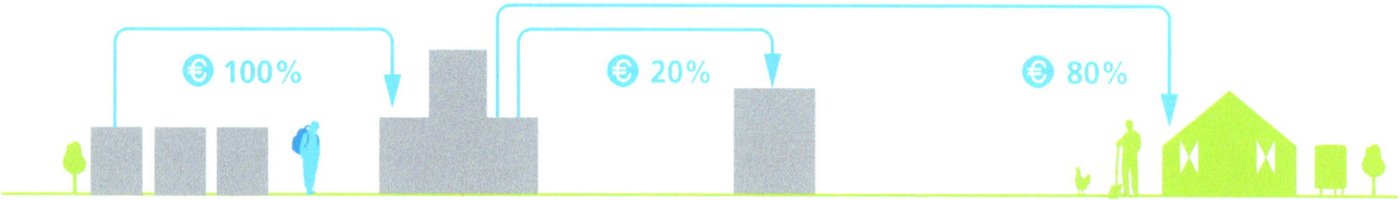

€ 100%
€ 20%
€ 80%

Landzijde turnover from care activities:

'06	1,1 m
'07	1,9 m
'08	4 m

€4,657
per year is the average cost per client

421
clients in 2009

101
average number of half-days per client

102 farmers were affiliated in 2009

Income per farmer:
€3,000 - €180,000
(average €70,000)

Professional, well-organised care and daytime activities

Preservation of the agricultural cultural landscape

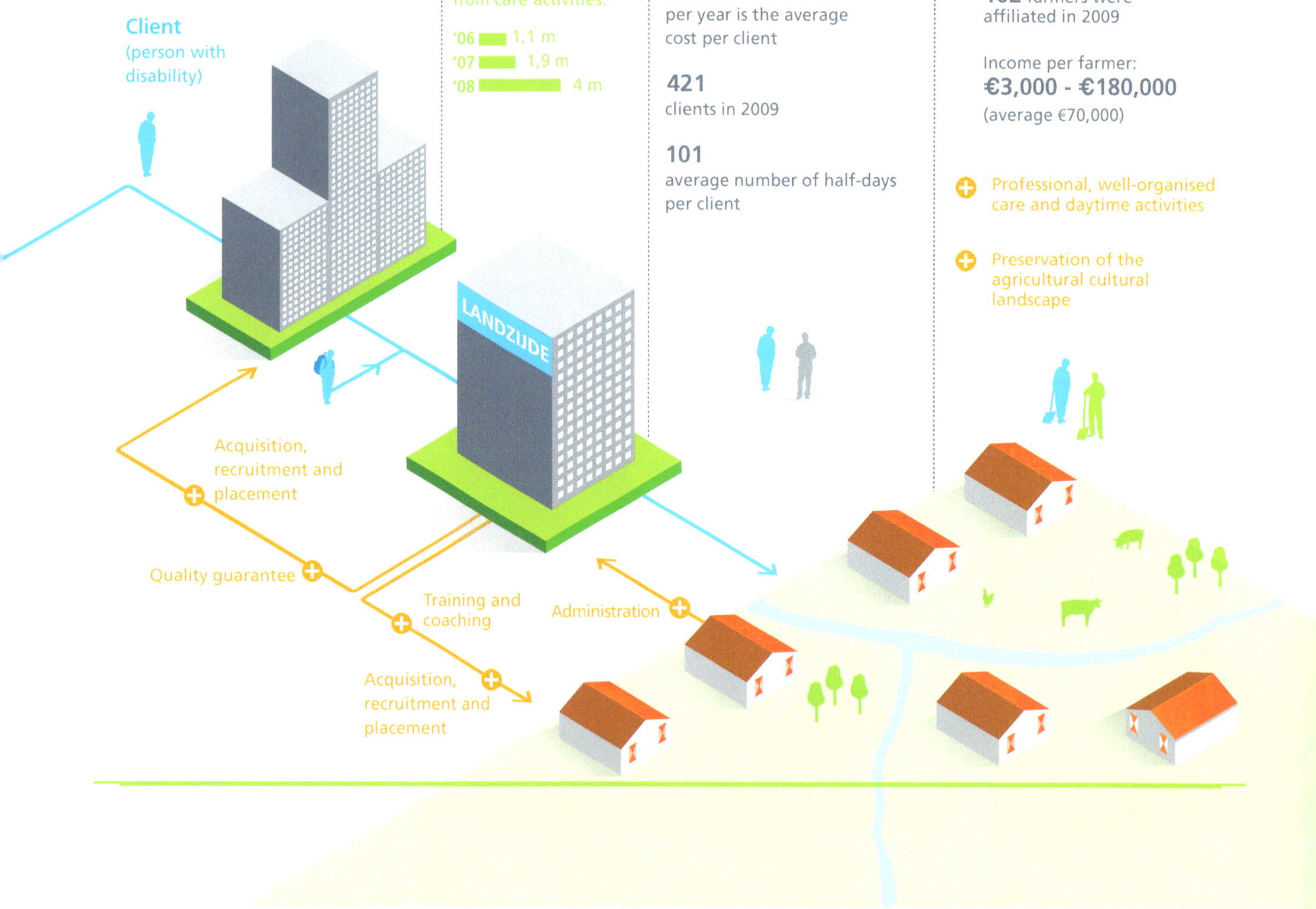

Client
(person with disability)
LANDZIJDE
Acquisition, recruitment and placement
Quality guarantee
Training and coaching
Administration
Acquisition, recruitment and placement

to determine which target group would be right for the farmer and whether the farm is suitable. Landzijde arranges the necessary permits and insurance, supports the farmer and provides training and courses. This provides a quality guarantee for clients, care institutes and financiers and helps prevent misunderstandings. On account of its scale, professionalism and links with research institutes, Landzijde has obtained medical, social and therapeutic recognition.

2. A single point of contact and a wide choice

The client or care institute seeking a suitable care farm can indicate its wishes on an intake form and cast an eye over the available care farmers online. The most suitable and desirable care farm is then jointly determined in an intake interview. This spares the client from the need to identify and assess all kinds of different care farms in the region. The care institutes have access via a single point of contact to tens of suitable farms, each concerned with specific target groups. A number of care organisations (for example the Municipality of Amsterdam and the Care Office) do not make appointments with individual care farmers.

3. Support for farmers

The administration required for care farmers is extensive and is still growing. Landzijde is able to take this over from the farmer, to arrange the necessary insurance and to provide support in respect of permits and zoning plan procedures. They also advise farmers on the modifications that need to be made to the farm.

4. Acquisition, recruitment and placement

The recruitment of new clients and care institutes is undertaken by Landzijde, not by the individual farmers. This generates efficiency gains, so that the acquisition is more professional and has a wider reach.

5. Education and courses

Landzijde believes in continuous professionalisation. Education and courses are held in order to train new farmers and provide additional training for experienced care farmers.

Courses are also provided for the staff of care institutes, often in conjunction with the care farmers. Intervision is organised for the care farmers. Landzijde collaborates with the Groenhorst College, in the form of a secondary vocational education course in care farm management.

6. An effective and influential discussion partner

On account of its size and network Landzijde acts as a discussion partner for municipalities, care insurers and care institutes. It also has access to the provincial government and ministries. This enables Landzijde to advise these institutions on the development of new policies and to mediate and advise on behalf of the affiliated care institutes and care farmers.

The sustainability performances

People

Generally speaking clients indicate that they are satisfied with the care farms and feel comfortable in their work. In this way working on the care farms contributes to clients' well-being. No hard information has, however, been found in order to determine whether clients' health does in fact improve from being on the care farms.

Apart from their 'ordinary' work on the farm, the farmers are required to devote attention to the clients. It is not known whether this leads to substantially more work for a care farmer as compared with an ordinary farmer. Many of the farmers in the scheme go in for landscape management, thereby helping to foster the kind of landscape desired by urban dwellers.

Planet

The primary objective of care farmers is care, and not necessarily more environmentally-friendly farming. Around 10% of the Landzijde farms are organic, as compared with 2% for the Netherlands as a whole. This suggests a positive relationship between care farming and organic farming.

No research is however available to demonstrate that care farmers farm more sustainably, extensively or organically than mainstream farmers. Apart from that no data were available on the planet performance of care farmers.

Profit

Landzijde is a non-profit organisation. The profit is spent on staff, accommodation, management, improving the standard of service and the development of new products. This allows Landzijde to offer its care at 7% below the normal price. The farmer receives 80% of the sum paid to Landzijde. Of the 20% in overheads, 12% is directly related to client care, such as intake, support and evaluation. Landzijde has been in the black in recent years.

Care farmers receive an average of € 55 per client per day. Generally speaking care farmers have more employees than ordinary farmers. The farmer's partner often plays an important role in supporting the clients. In many cases she will have a background in care.

Additional costs faced by care farmers are:
- Food & drink during the day
- Extra staff
- Training
- Regular quality control
- Costs of extra activities for clients
- Administration

On average six to seven clients will work on most Landzijde farms. The necessary investments the farm has to make are comparatively small and not a factor of significance. An average care farmer affiliated to Landzijde earns approximately € 70,000 a year from care.

SWOT analysis of the sustainability performances

Strengths
- Efficient value creation with a new product: relatively more value per investment in labour is created than for example in the case of a market launch of new food products by farmers.
- Positive effect on the client's budget (7% in relation to mainstream care).
- The business case can be scaled up and is transferable to other regions (also an opportunity).

Weaknesses
- Many extra transport movements.
- Risk of transmission of animal diseases to human beings in the case of an epidemic.

Opportunities
- Encouraging care farmers to switch to organic methods.
- Encouraging farmers to take up meadow bird management and other forms of nature management.
- Developing relationship with primary schools concerning green education.
- Strengthening regional development by lobbying for more bus routes and stops.
- Research into the actual effects of care farming on physical and mental health.

Threats/risks
- Dependability of government policy/AWBZ (General Exceptional Medical Expenses Act), which can be subject to ad hoc cuts.
- Outbreak of animal disease epidemics can lead to the temporary or permanent closure of farms.

8.5 The value creation model

Notes on the value creation model

The value creation model describes how Landzijde creates value for all the stakeholders and invests the proceeds in the further professionalisation of the industry. Landzijde provides the farm with added value as a setting for 'recovery'. The ever increasing professionalisation sets Landzijde apart from other initiatives, which in turn generates higher results.

The creation of values may be shown as follows in the model (see below).

8.6 From plan to investment

Landzijde was established in 2000 and officially recognised in 2003 as an AWBZ care institution. The investment costs in order to set up this care institute during the period 2000-2003 amounted to approximately € 300,000 (source: J. Hassink, Business models in green care).

These costs were composed as follows (amounts in 2003):

- Acquisition, talks with care institutes, etc: € 90,000.
- Setting up the administrative and financial system: € 90,000.
- Leaflets, brochures: € 40,000.
- Setting up the office: € 40,000.
- Support for first 50 farms in the form of first aid course, clothing, etc: € 40,000.

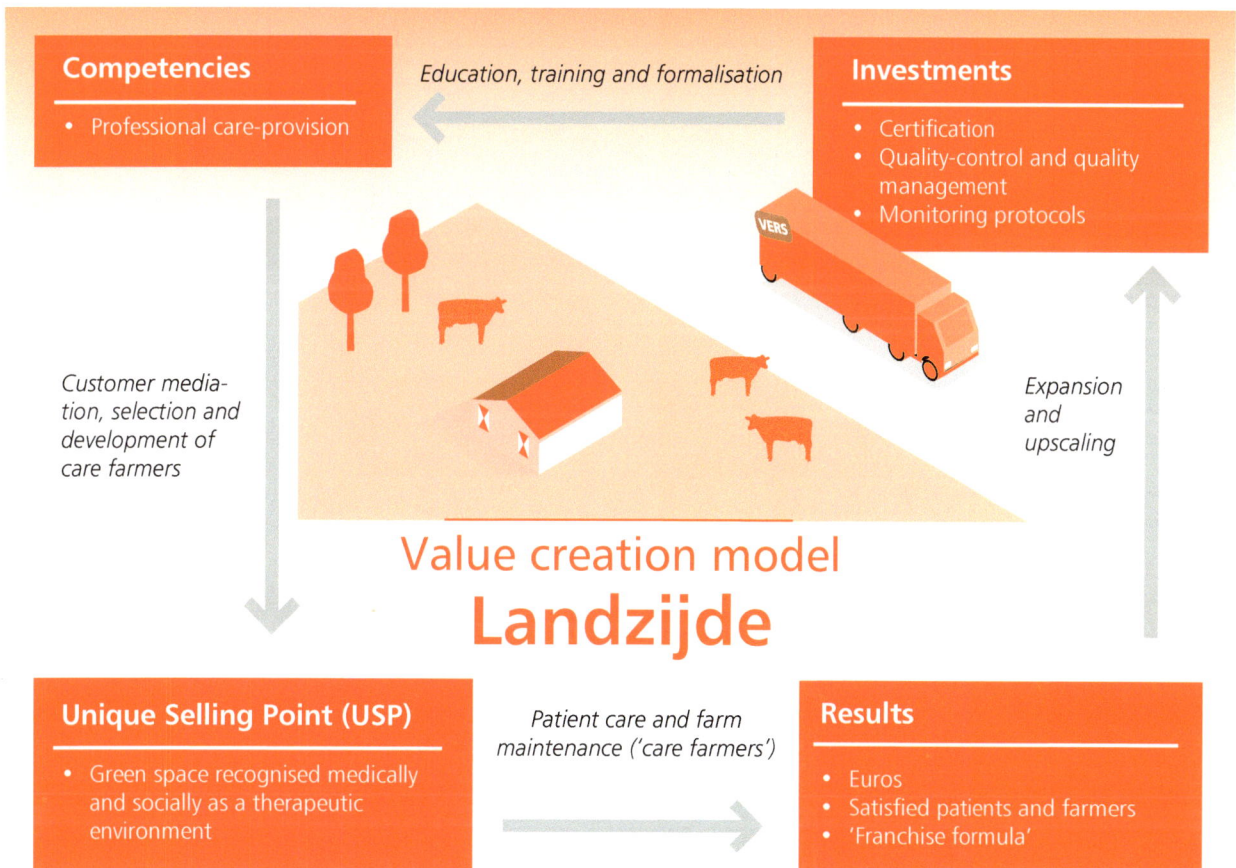

Competencies
- Professional care-provision

Education, training and formalisation

Investments
- Certification
- Quality-control and quality management
- Monitoring protocols

Customer mediation, selection and development of care farmers

Expansion and upscaling

Value creation model
Landzijde

Unique Selling Point (USP)
- Green space recognised medically and socially as a therapeutic environment

Patient care and farm maintenance ('care farmers')

Results
- Euros
- Satisfied patients and farmers
- 'Franchise formula'

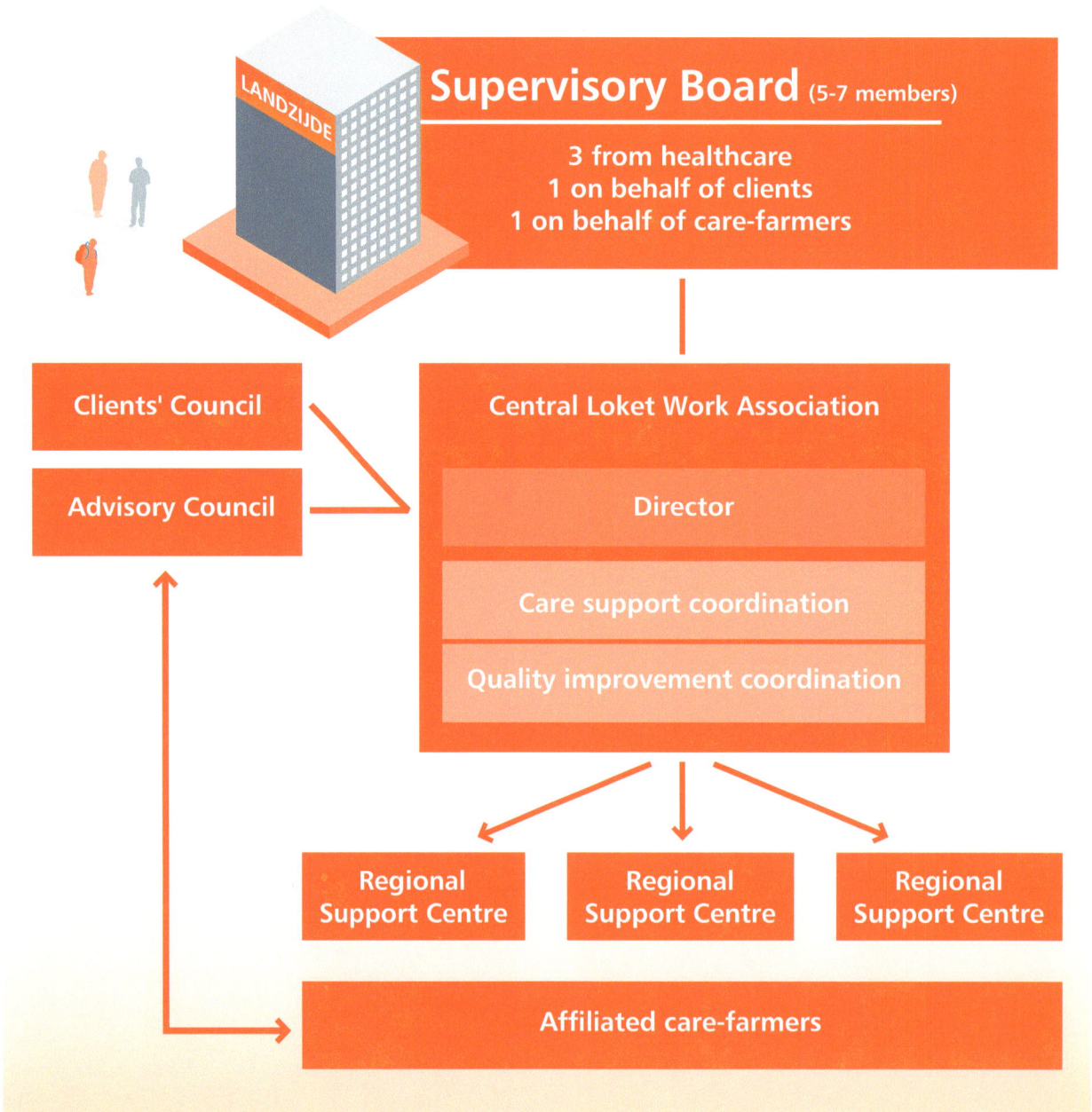

LANDZIJDE

Supervisory Board (5-7 members)

3 from healthcare
1 on behalf of clients
1 on behalf of care-farmers

Clients' Council

Advisory Council

Central Loket Work Association

Director

Care support coordination

Quality improvement coordination

Regional Support Centre

Regional Support Centre

Regional Support Centre

Affiliated care-farmers

The costs were partly covered by subsidies from the Province of North Holland (€ 110,000), the Municipality of Amsterdam (€ 25,000) and Doen Foundation (€ 40,000). In addition the founder invested a year or so of his own time. After the AWBZ-recognised Landzijde had 50 care farmers in 2003, the province provided an additional € 270,000 to fund further expansion. Since 2005 Landzijde has had a positive cash flow.

The organisation

The Landzijde organisational model indicates that it has embedded various chain players (from care and agriculture) and its clients in its organisation, through both the Supervisory Board and the advisory councils.

8.7 The lessons for the entrepreneur

The entrepreneur as project developer

The various stakeholders changed role in the planning, investment and operating phases. The future 'customers' and financiers (care institutes, care insurers, the municipality and the province) were closely involved in the development of the business model. During the planning phase consultations were held with these stakeholders, who formulated their wishes and contributed ideas towards the setting up of Landzijde. In the investment and realisation phases this relationship became more 'business-like', as the roles of client, supervisor and financier are clear. Switching from one role to another in existing relationships is not always straightforward and is best discussed openly.

The entrepreneur as coach

The organisation of workshops, in which employees of care institute and care farmers considered and discussed the establishment of Landzijde, made an important contribution towards the design and realisation of the 3P business case.

The entrepreneur as strategist

Value creation requires a professional care farming sector that links up care and agriculture. Professionalisation is essential here, as Jaap Hoek Spaans understood like no other. In order to guarantee continuity as well as the necessary quality, care farming needs to be taken out of the domain of subsidies and (well-intentioned) idealism. Subsidies are dependent on (inconsistent) government policy, and unprofessional conduct can result in abuses that harm vulnerable clients. Furthermore serving the metropolitan market means operating on a certain scale, which calls for a professional organisation.

The entrepreneur as games-leader

Allow yourself as entrepreneur to have your own strengths and weaknesses identified (for example by organising external reflection) and compensate for the weaknesses you bring to the project by appointing people with complementary competencies.

Be flexible in the development of methods, do not cling onto methods and instruments that were devised at the start of the process and provide space for new questions and topics that emerge as matters unfold.

The entrepreneur as spider in the web

New insights arose from cooperation with research bodies and other institutes and established organisations and from the continuous integration and supervision of participating farmers and care institutes. The director of Landzijde devoted a lot of time and energy to communication and the development and management of the local, regional and national Landzijde network.

In terms of supply and demand, Landzijde plays the role of central chain orchestrator, linking up demand (care) and supply (care farming). Farmers, clients and care institutes all have a place in the organisational structure.

The entrepreneur as winner

The professionalisation of care farming means taking (market) demand as the starting point. The demand and the needs of care institutions and clients therefore act as the point of departure for the development and management of chains linking up the care demanders and the providers (care farmers).

This calls for a chain orchestrator who is familiar with and able to connect up the two worlds. By creating added value for all concerned, it became possible to develop a business model that did not depend on subsidies. Is therefore necessary to work continually at improving the supply and to respond alertly to any new (financial) opportunities that may arise.

Changes in the care financing system have had a positive impact on the development of care farming. Since the introduction of the Personal Care Budget (PGB) the growth of care farms has gone into higher gear. After fluctuating around a figure of around 350 for several years (2001-2003), the number of care farms rose to 850 in 2008.

Independent quality assurance is important for the development of new product, particularly when this concerns a service. The 'quality label' at Landzijde was the AWBZ recognition – a precondition for receiving insurance-funded clients. Quality assurance and monitoring and AWBZ recognition are important for the client, care institutes and insurers.

8.8 The present challenges

Scaling up

Landzijde now operates as a professional provider of care farming. After the closure of the National Agriculture and Care Support Centre, a national debate got under way as to how the care farming sector should develop. Given its market share the sector has the potential to develop into an important economic player in the metropolitan area:

urban demand and rural supply are being linked up in new value chains. Landzijde currently participates fully in this debate, stressing the need for professionalisation and the importance of taking care rather than agriculture as the starting point in the development of new business plans. Landzijde now faces the choice as to whether it wants to continue working regionally, to extend its area of operation or to work on a franchise basis.

Towards independent certification and quality control

A fully-fledged sector also means the development of new national certification systems and the organisation of quality rules, licensing systems and supervisory arrangements. All this is still in its infancy in the Netherlands, meaning that there can also be anomalies. These can be a source of difficulty for professional concerns such as Landzijde. The organisation would therefore benefit from the independent regulation of standards, similar to those for childcare.

Broadening out to new markets

Experiments have commenced within the TransForum project into combinations of special education and care in a farm setting. The initial experiments with the farm as a learning environment for children with learning and behavioural difficulties appear highly promising. Children in special education spent several half days a week on the farm. The results showed that they learned better and developed competencies that were much less evident in a traditional classroom setting, such as self-confidence, co-operation and taking responsibility. Professional chains still need to be developed for this. Landzijde can play a role here.

National/international marketing of know-how

As a pioneer, Landzijde has built up a wealth of knowledge and experience. The obvious course of action would be for these to be commercialised. The Landzijde business model would appear capable of being applied internationally as well if translated for the specific context of a particular country. The cultural circumstances, financial flows and organisation of care and agriculture differ from country to country. Steps are currently being taken in Brazil to translate the this experience to a Brazilian 'Green Care'.

Jaap Hoek Spaans, managing director of Landzijde

(Photo: Mugmedia, Wageningen)

The farmers in the NFW manage 1,700 km of wooded banks and alder belts. (Photo: Mugmedia, Wageningen)

9. NORTHERN FRIESIAN WOODS

farmers jointly create value

9.1 The challenge

Dutch society attaches great value to ancient cultural landscapes. *'The Northern Friesian Woods'* area, located in the north of the Netherlands, is such a unique area. The farmers in the area saw this as an opportunity. They had already built up years of experience with managing the landscape themselves and dealing with the environmental regulations.

After years of meeting, consultation and research, the chairman of the Association of Northern Friesian Woods, *Douwe Hoogland*, posed the question: that's all well and good, but *'Wat smyt it up?'* – or, *'What's in it for us?'* In doing so he was articulating the real challenge: the development of new product-market combinations (PMCs) that would turn the National Landscape to commercial account for its entrepreneurs and citizens, and linking up the individual PMCs into a single business case for the commercialisation of the National Landscape in a way that makes use of all the parts. The farmers began to explore new markets and develop new products.

9.2 How did the innovation come about?

The mission of the Association of Northern Friesian Woods

'The development of agriculture and the regional economy, in conjunction with strengthening the cultural-historical landscape and its ecological features.'

Self-organisation and self-steering ('We can do and will do it better')

The most important driver for the entrepreneurs in the area is that they want to work in their own way at the landscape, environment and nature on their farms. They are convinced that this will generate better results for the area and their farms than continuing with the present practices. Present

The vision of the Association of Northern Friesian Woods

'Taking the responsibility for production, market development and the landscape and sustainability of the region.'

practices are directed by government through prescription of 'resources and instruments'. The farmers in the Northern Friesian Woods want to work towards a system that is geared towards targets: something that does not sit easily with provincial, national and European legislation.

In 2002 six agricultural nature conservation associations and environmental cooperatives consequently joined forces and set up the Association of Northern Friesian Woods. This association now represents some 850 farmers with more than 40,000 ha of land. Together with research groups from Wageningen University and the province they established working groups and study groups focusing in particular on the provision of support for farmers in landscape management.

An important next step was to draw up a programme of work and the signature in 2005 of an agreement with all the main players in the region (government authorities, NGOs, entrepreneurs and also Wageningen University).

Project partners
Centre for Agriculture and Environment, Netherlands Energy Centre, Institute for Agricultural Law, LTO North Projects, Groningen UR, TNO, Association of Northern Friesian Woods, VU University Amsterdam (Athena Institute), Wageningen UR (Alterra, Rural Sociology, Public Administration) and TransForum.

TransForum project
2004-2010

In order ultimately to achieve self-steering, the association wants:

- To agree at product level on the kind of landscape and environmental quality to be produced by the entrepreneurs.
- They themselves determine how they will do this: for this purpose they have set up a review committee to advise and supervise the farmers.
- Concluding agreements in respect not just of the landscape but also of the environment and water.

The farmers also had their own wishes with regard to their business operations. In 2001 60 livestock farmers together with researchers from Wageningen UR set up a minerals project. The project was designed to improve soil fertility and soil biology by increasing the quality of manure and reducing the emission of minerals by the use of low-protein food.

Part of these business operations consisted of the surface-spreading of manure, which ran up against the existing environmental regulations, which prescribe the injection of manure. Ministers of Agriculture Veerman (2006) and Verburg (2010) provided dispensation from the manure regulations for this purpose.

When the area was designated as National Landscape in 2006 the farmers remained in the driving seat and became the most important pioneers of the National Landscape.

Forging alliances ('Not by oneself')

In order to achieve their goals, the entrepreneurs not only organised themselves but also developed networks inside and outside the area. Within the area they forged coalitions with municipalities, the province, societal organisations and other local entrepreneurs.

They also invested heavily in their relationships with the Ministries of Agriculture, Nature and Food Quality (LNV) and Housing, Spatial Planning and the Environment (VROM) and with politicians from the region or who were sympathetic to their aims.

Developing business cases ('What's in it for us?')

One drawback was that the association got caught up in consultation, lobbying and research. The association therefore decided that it also wanted to work on value creation for the region and the farmers in particular. Together with researchers and civil servants they conducted a survey into potentially successful product-market combinations.

Among other things this study focused on the potential 'metropolitan' market for new products. This was followed up by a regional meeting at which the new products and methods of production that were the most important for the region and that released the most energy were selected.

Pingo ruins: valuable relics from the Ice Age

New methods of production and new products

1. **Regional branding:** Uniform promotion of the area, the Recreation & Tourism development organisation and regional products.
2. **Energy from wood:** Wood traditionally served a utilitarian function. Now there is a restriction, in that a use has to be found for wood as a waste substance. Waste wood becomes valuable again.
3. **Remote control:** Regional responsibility for the implementation of the new agricultural nature management subsidy scheme (SAN).
4. **Harmonisation of agriculture and the landscape:** setting up a modern dairy farm within the National Landscape.
5. **Landscape vision and design:** coordination and standardisation of policy for the Northern Woods National Landscape and inventorisation and management of landscape elements.
6. **Closed-loop farming:** milk obtains extra value when linked to sustainability (closing regional loops) and the landscape.
7. **Clean water through self-steering:** an improvement drive in the quality of surface waters, e.g. by drawing up a region-specific yardstick and monitoring system.

A single story and a single business case for the entire region

A strong, publicly backed 'regional story' provides the foundation for the policy of 'making use of all the parts' of an area: it links the various product-market combinations and motivates the various stakeholders in the area. The 'regional story' needs to be laid down in binding agreements so as to prevent brands from becoming contaminated or everyone developing their own brand.

The individual business cases of the product-market combinations are amalgamated into a single business case describing how the National Landscape as a whole can be turned to commercial account by the combined entrepreneurs in the area. To this end a transfer and equalisation mechanism was developed.

Inspiration from outside

New ideas arose from discussions with researchers and practical experts and from visits to other areas, such as the Groene Woud ('Green Wood') in the Province of North Brabant, where a financial transfer mechanism was introduced based around an 'regional savings account'.

Dealing with the bureaucracy

In order to turn the integral Northern Friesian Woods business case into a success, the association must be able to manage this itself. The existing structures, regulations and bureaucracy were a significant obstacle for the association. Together with researchers and 'enlightened' civil servants, the entrepreneurs tried to find a response to the commonly voiced expression, 'Brussels won't allow this'. The researchers analysed the policies and regulations of the EU and the Netherlands and collectively got in touch with European and Dutch civil servants and politicians.

Strong public support and direct contact with national, provincial and municipal politicians and administrators created the space for the necessary dispensations and room to experiment. Experiments were substantiated and monitored by researchers.

The political lobbying was conducted with the aid of MPs from the region and influential academics. The ultimate goal of all these efforts was to restructure the generic (general) regulations and generic policies into more customised, area-specific policies based around closed-loop farming and landscape management. So far this has succeeded on an experimental basis, but it is not yet become embedded in policy or legislation.

2001	Start of minerals project by farmers and Wageningen University
2002	Establishment of the Association of Northern Friesian Woods by six agricultural nature associations
2004	Start of TransForum Northern Friesian Woods project
2005	Signature of Northern Friesian Woods declaration of intent and working programme by regional stakeholders, ministries and Wageningen University
2005	Farmers' study groups set up
2006	Northern Friesian Woods become part of the new 'Northern Woods' National Landscape
2006	The Dutch Green-Blue Services Catalogue based on the experience of the Northern Friesian Woods is completed and is sent by the Ministry of Agriculture, Nature and Food Quality to the EU
2006	Motion in the national Parliament calls for room for an alternative track (including the surface spreading of manure)
2006	Minister of Agriculture Veerman provides dispensation for manure surface spreading
2008	Start of pilot with own review committee as part of landscape management programme
2008	TransForum Self-Steering and Profit Project
2010	Northern Friesian Woods Self-Steering and Profit Pilot in the national Parliament under Agenda on Vital rural areas and manure policy
2010	Minister of Agriculture Verburg provides space to experiment with 'Closed-loop farming'

9.3 Key figures

- Association of Northern Friesian Woods: 850 farmers with 40,000 ha of land.
- Area covered by Association of Northern Friesian Woods: 148,000 inhabitants.
- Energy from wood: the farmers in the Northern Friesian Woods manage 1,700 kilometres of wooded banks. This can result in an annual yield of 10,000 tonnes of pressed wood pellets.
- Milk: the farmers in the Northern Friesian Woods produce some 300 million litres of milk a year, or 2.7 % of annual Dutch milk production. The milk is delivered to dairy company FrieslandCampina. At a price of € 0.31 per litre (2009) this generates income of 91 million euros.
- Nature and landscape management: 10,500 ha of meadow bird management, 4,000 ha of foraging land for goose, management of 150 km of wooded banks and 1,500 km of alder belts, hundreds of pingos and bog holes and 800 ha of botanical nature management.

9.4 The added value of the Northern Friesian Woods

The advantages of the Northern Friesian Woods

1. Quality and efficiency gains from self-organisation.
2. Clear-cut public perception of the region ('one brand').
3. Equalisation between the regions and reduction in dependence on subsidies.
4. Products that tie in with (metropolitan) market demand.

1. Higher quality and lower costs

Giving the farmers greater responsibility increases their involvement and motivation. It also taps the regional expertise and inventiveness. A reduction in bureaucratic regulations, supervision and enforcement cuts down the costs of policy implementation.

The Northern Friesian Woods

New product-market combinations between rural areas and the urban market are being developed for the Northern Friesian Woods.

⊕ The advantages

Regional branding

Promotion of the area, Recreation & Tourism development organisation and regional products.

Landscape vision & design

Harmonisation of agriculture with the National Landscape and management of landscape elements.

Added value from milk

Milk obtains added value when linked to sustainability and the landscape: closed-loop farming.

Energy from wood

Wood becomes a resource again.

Clean water through self-steering

Improvement in the quality of surface waters, e.g. by means of region-specific yardstick and monitoring.

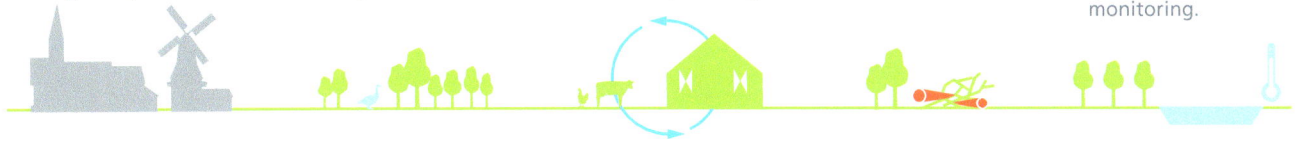

⊕ Quality and efficiency improvements through self-organisation

⊕ Regional levelling and reduction in dependence on subsidies

⊕ Products that correspond with (metropolitan) market demand

1,700 km total length of wooded banks

⊕ Distinctive regional image ('single brand')

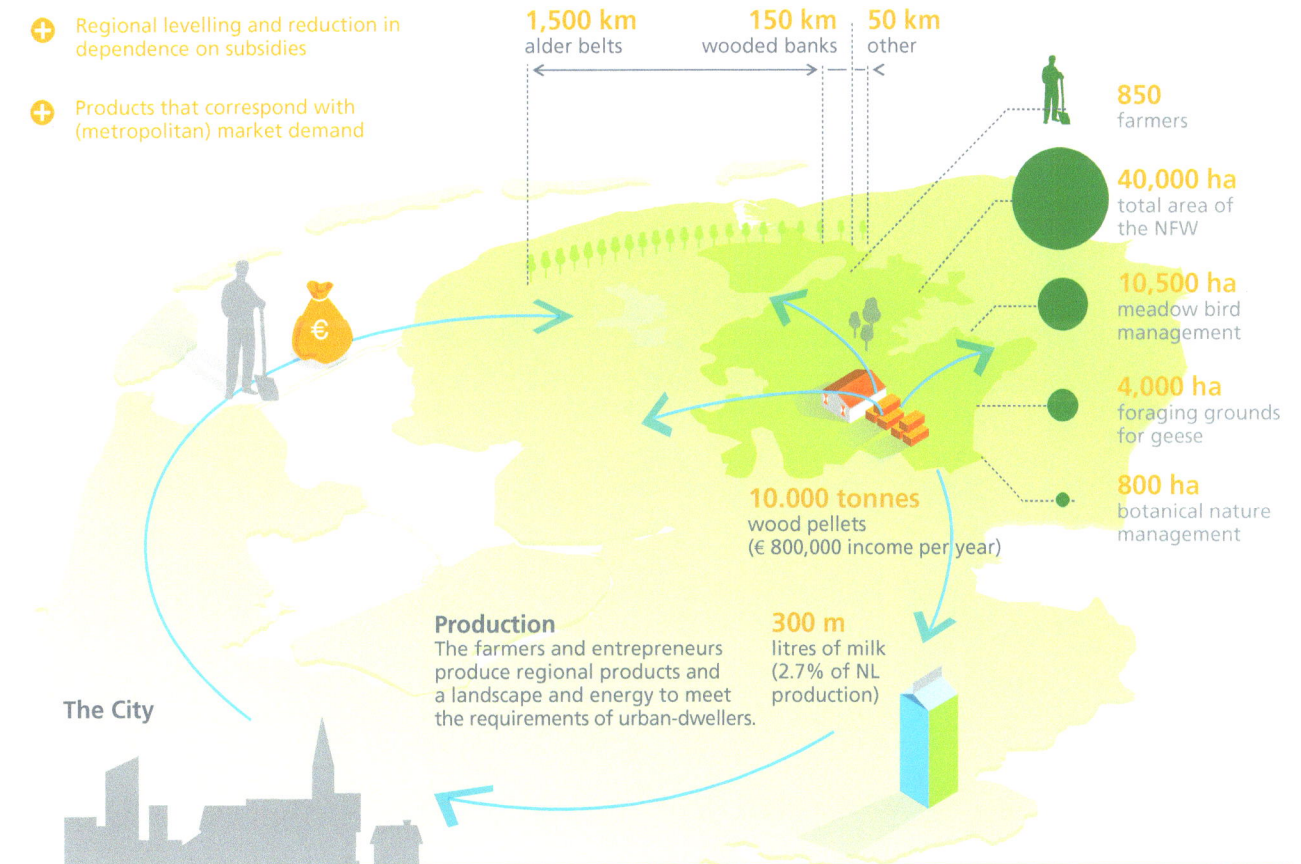

1,500 km alder belts **150 km** wooded banks **50 km** other

850 farmers

40,000 ha total area of the NFW

10,500 ha meadow bird management

4,000 ha foraging grounds for geese

800 ha botanical nature management

10.000 tonnes wood pellets (€ 800,000 income per year)

The City

Production

The farmers and entrepreneurs produce regional products and a landscape and energy to meet the requirements of urban-dwellers.

300 m litres of milk (2.7% of NL production)

The storybook atmosphere is valued by urban dwellers (Photo: Mugmedia, Wageningen)

2. The Northern Friesian Woods as a single brand

As a result of the mutual cooperation and coordination among farmers, recreational and catering industry entrepreneurs, the municipalities and the province, a single story for the area is drawn up and a single area brand is developed. This helps support the marketing of the area and the local products, as it strengthens the recognition of the area at national and regional level. A shared story supports and encourages mutual cooperation within the area and strengthens the sense of 'pride'.

3. A single business case for the entire area

The individual product-market combinations are linked up in a single business case. This allows the one product-market combination to contribute to the other; the hotel and recreation industry in the area, for example, benefits from the existence of an attractive landscape and public footpaths, as provided by the farmers, and the income accrues primarily to that sector. An equalisation mechanism, such as a regional savings account makes it possible to transfer costs and earnings. The fact that landscape development is wholly or partly financed from such income makes the area less dependent on (uncertain) government subsidies.

4. New chains linking up city and countryside

The Northern Friesian Woods project has enabled new product-market combinations to be established between rural areas and the urban market. The farmers and other entrepreneurs in the area produce regional products, landscape quality and energy that meet the needs of 'urban dwellers'. In this way they create new value for the urban dweller, but also for themselves, making it possible to continue developing their farms as part of the National Landscape. This does however call for a new way of looking at things and new competencies, i.e. a switch from product-oriented to market-oriented thinking.

The sustainability performances

People
Landscape

By the active management and development of the wooded banks in the area, the farmers help enhance the publicly valued cultural-historical landscape.

Animal welfare

The Association of Northern Friesian Woods contributes to animal welfare since grazing by cows forms a essential element of the 'closed-loop farming', as laid down in the Woods Certificate ('Woudencertificaat'). In this way the Association of Northern Friesian Woods generates added value in relation to normal farming practice, where cows are increasingly kept in the shed.

Social cohesion

The regional branding based around the shared 'Northern Friesian Woods story' means that joint initiatives are supported by non-agricultural entrepreneurs and that citizens become involved with the area. All this helps strengthen the local community.

Health

The CLA-content (unsaturated fatty acids, such as Omega-3) in the milk from the Northern Friesian Woods turned out to be higher than elsewhere, which may be due to the fact that the cows feed more on grass than maize silage during the winter period. This may be one of the causes for the healthier milk from the region.

Employment

Most of the employment in the Northern Friesian Woods is generated by the agricultural sector. Strengthening the economic position of that sector helps maintain employment in the region.

Planet

Reduction in nitrogen and ammonia emissions

The system of closed-loop farming in the Northern Friesian Woods seeks wherever possible to develop agriculture as a cyclical system. Roughage is cultivated within the region, while the cow manure is spread out on the land. This results in high protein food for the cows, leading in turn to lower nitrogen and ammonia emissions into the atmosphere.

Biodiversity

In order to reduce the nitrogen content the grass is mown later, which is better for meadow birds. Meadow bird management forms part of the system of circular agriculture. The wooded banks in the area are managed in collaboration with biologists with a view to preserving and enhancing biodiversity.

Profit

The economic development in the area is strengthened as the activities of the Association of Northern Friesian Woods can boost the number of recreationists visiting the area and lead to new economic initiatives. This can give rise to investments in infrastructure and public transport. Farmers' incomes in the Northern Friesian Woods are currently low. Virtually all the agriculture in the area is concerned with dairy farming. The abolition of the milk quota in 2015 by the European Union, will lead to lower milk prices. In order to be successful, the new business models accordingly build on the current practice of dairy production in a small-scale landscape. Higher earnings could be obtained from milk if added value could be given to the milk by linking the product to landscape management and closed-loop farming and by emphasising the health benefits of the local milk or 'Woudenmelk'. The landscape management would be commercialised in the form of the payments made to farmers by society for this purpose and by extracting energy from timber.

SWOT analysis of the sustainability performances

Strengths

- Close relationship between farmers and citizens in the area, contribution towards regional development.
- Landscape management as an integral part of farming operations.

Weaknesses

- Closed-loop farming is uncertain: this is now possible thanks to the numerically and time-limited experiment, which is holding back investment and upscaling.

Opportunities

- Investing in the business cases will lead to an increase in the level of economic activity and investment in the area (generating a knock-on effect).
- Community formation through the link-up of farmers/ other entrepreneurs/citizens/municipalities.
- Make use of the higher CLA-content in milk in product marketing.

Threats/risks

- A future worsening of the economic circumstances of dairy farming and the falling number of farmers could lead to a deterioration in the agricultural cultural landscape.

9.5 The value creation model

Notes on the value creation model

The unique feature of the Northern Friesian Woods is the small-scale agricultural cultural landscape. The business case for the area is based around the commercialisation of this unique selling point. To this end the entrepreneurs (farmers and entrepreneurs in the hospitality industry) want to co-operate closely with local government in order to promote the area. The farmers are developing a business system of sustainable dairy farming in a small-scale landscape, known as closed-loop or circular farming. They hoped to increase their earnings from special milk,

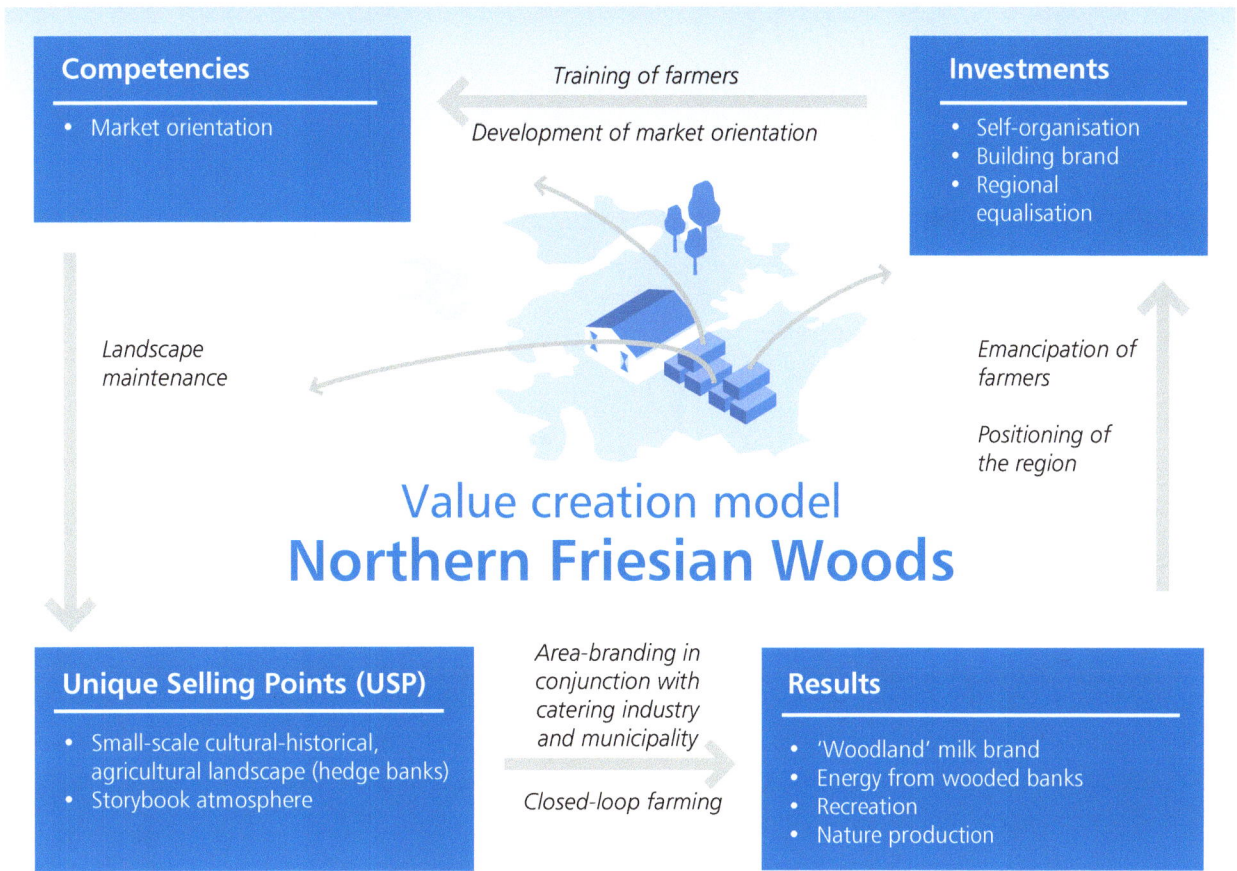

Competencies
- Market orientation

Training of farmers
Development of market orientation

Investments
- Self-organisation
- Building brand
- Regional equalisation

Landscape maintenance

Emancipation of farmers

Positioning of the region

Value creation model
Northern Friesian Woods

Unique Selling Points (USP)
- Small-scale cultural-historical, agricultural landscape (hedge banks)
- Storybook atmosphere

Area-branding in conjunction with catering industry and municipality

Closed-loop farming

Results
- 'Woodland' milk brand
- Energy from wooded banks
- Recreation
- Nature production

---

energy, nature and recreation by developing a strategy for the entire area. This requires investment in a strong organisation and a common brand. Since the income and expenses in this business case do not always relate to the same entrepreneur, some system of equalisation is required. Among other things a 'regional savings account ', modelled on the Green Wood initiative in the Province of North Brabant, is being studied for this purpose. The creation of values may be shown as follows in the model (see above).

9.6 From plan to investment

The Association of Northern Friesian Woods has not yet reached the point of actual investment. Following an analysis of the region and a survey of promising product-market combinations, working groups of civil servants, farmers, researchers and other practical experts are at present writing up business cases for the selected combinations. Nyenrode Business University has analysed a number of the selected business cases and estimated the potential earnings.

Added value from Milk
On the basis of total milk production of 300 million litres and a milk price of € 0.31 a litre, an 8% increase in the price of milk could lead to an increase in average farm income of € 8,700 a year. This would however mean that the dairy processor would have to be prepared to market the milk as a 'specialty' (i.e. regional milk).

Landscape management
Via the EU and national and provincial governments, society at present pays one euro for each m^2 of wooded bank. The farmers in the Northern Friesian Woods currently manage 1,700 kilometres of wooded bank, generating a cash flow of 1.7 million euros. Depending on the area of wooded bank on the farm, this amounts to an annual increase in farmers' income of some 2,000 euros. The Association of Northern Friesian Woods is aiming at an agreement based on goals and output instead of resources. This could mean a substantial savings for the government in terms of inspection and enforcement costs.

Energy from wood

At present the farmers are obliged to take their wood trimmings to waste dumps. This costs 20 euros per tonne. Processing the wood into wood pellets for burning would result in net revenues of around 800,000 euros, or an average of 941 euros per farmer per year.

Closed-loop farming

Potential cost savings on manure processing of around 3,750 euros per farm are possible in comparison with the traditional method of high-protein feed and manure injection.

Other sources of income cited by Nyenrode Business University are the production of cheese (high earnings, but also high costs), solar panels (dependent on energy subsidies), care farms and childcare (for a limited number of farms) and the development of recreational facilities (including accommodation) in the area.

The latter would however require a transfer mechanism whereby part of the earnings received by recreational entrepreneurs from this source would be transferred to the farmers for the management of the area and the development of, for example, footpaths and cycle networks. The instrument of a 'regional savings account' that has operated so successfully in the Green Wood initiative in the Province of North Brabant is an example of such a transfer mechanism. In the Province of North Brabant entrepreneurs, government authorities and private individuals help invest in the area via this account.

9.7 The lessons for the entrepreneur

The entrepreneur as project developer

Investments to get maximum value out of all parts of a National Landscape are complicated, as these are both public and private investments. The private investors invest in business plans involving recreation, energy and dairy products. Public investors invest in public goods such as nature and the landscape. During the planning phase, a 'mixed' business case must demonstrate the potential for value creation in the area.

The entrepreneur as coach

Room to experiment was an essential condition for the developments in the Northern Friesian Woods. The province and national government provided such room for the development of closed-loop farming and the landscape. The region accordingly became an important 'trial–plot' for the government to experiment with self-steering in a National Landscape. This in turn led to the willingness of and potential for politicians and civil servants to contribute their ideas and to cooperate.

The entrepreneur as strategist

The designation of the area as National Landscape was picked up by the farmers as an opportunity for the development of new markets. City dwellers attach great value to cultural-historical landscapes that embody their romantic image of the 'traditional Dutch landscape'. This provided an opportunity to develop new business cases. The entrepreneurs in this area have concentrated in particular on the strategy of 'Sustainable Diversification': by responding to the new requirements of city dwellers, agricultural entrepreneurs are able to develop new products and markets.

The entrepreneur as games-leader

Entrepreneurs need to keep hold of the reins right from the start and to reach clear agreement if other players are not to dominate the process. The entrepreneur who cooperates with all sorts of parties soon discovers that government authorities and researchers, etc, all have their own rules of the game. This means that they go off at a tangent, postpone choices or spill out over onto different topics.

Clear agreements therefore need to be reached with one another as to the expected end-product. Try to reach agreements with government authorities at 'product level', for example concerning the type of landscape they consider important, instead of at 'resources' level, for example how wooded banks are to be maintained.

It is also important to assess from the outset which activities will generate the most income. Some activities call for a great deal of energy or look highly attractive, but are not necessarily those that ultimately lead to the highest earnings.

The entrepreneur as spider in the web

Organising the farmers in the region into a single association was an important success factor in the Northern Friesian Woods. It meant that they became an interesting partner for government authorities and research institutes and that they were able to develop new knowledge collectively. Bringing hundreds of farmers in an area together and reaching agreements with the government authorities requires someone who enjoys trust within the region and also at government level.

The agricultural entrepreneurs in the Northern Friesian Woods opted to collaborate with colleagues and to seek out allies who were able to help them realise their goals. The new products can only be brought to fruition through cooperation among the farmers themselves, between farmers and other entrepreneurs (e.g. in the recreation industry) and between farmers and the public sector and research institutes.

Over the years the Association of Northern Friesian Woods built up an influential regional and national network of knowledge institutes, government authorities, politicians, entrepreneurs and NGOs to drive forward the goals of knowledge development, public support and lobbying. Finally the new connections had to be formalised: the innovations were embedded in organisations, fixed structures, procedures and agreements, as a result of which the Association was accepted at government level as a consultative partner.

A 'story' is a linking factor. A strong, publicly backed 'regional story' provides the foundation for the policy of 'making use of all the parts' of an area: internally it links up the various product-market combinations and motivates the stakeholders, while externally it is important for the branding of the region and as the basis for marketing.

The Association of Northern Friesian Woods opted in favour of:

1. Having a regional story that strongly appeals to the public.
2. Creating value by the development of business plans for product-market combinations and self-steering.
3. Embedding by organising investments in new product-market combinations and area contracts/protocols for self-steering with government authorities and farmers.

In developing new products the entrepreneurs in the Northern Friesian Woods learned from others by visiting other areas and involving experienced practitioners in the project. The Association of Northern Friesian Woods for example visited the Green Wood area in the Province of North Brabant and involved an experienced entrepreneur in the development of the 'Energy from wood' business case.

The entrepreneur as winner

The agricultural entrepreneurs did not regard the inevitable designation of the area as a National Landscape as a threat but as an opportunity for the development of (3P) business cases. They focused on people, planet and profit values so that they could continue investing in their farms and create public support for agriculture and because they regarded (and continue to regard) themselves as the 'owner' of the area where they live and work.

Most entrepreneurs lack the time and skills for dealing with bureaucracy, government policy and the protracted consultations and paperwork that this entails. The Association of Northern Friesian Woods placed these activities in the hands of a few individuals who were into that kind of thing or who organised external support for the purpose.

Entrepreneurs also need to organise the necessary proof for products that are marketed on the basis of a sustainability story. In the Northern Friesian Woods researchers from Wageningen University have been providing the necessary scientific data to support the closed-loop farming system. The health claim for the milk was investigated and biologists have been involved in the management of the wooded banks.

9.8 The present challenges

From 'commodity' products to 'specialties': creating added value

The farmers in the Northern Friesian Woods want to generate added value through landscape management, farming and the production of healthy milk. This must lead to higher incomes. In the recreational market there are opportunities for operating on a national scale. Whether this applies to dairy products is open to question; for the time being the market will be a regional one.

In order to realise these business cases, chains will need to be set up and partners found in the logistical, processing and retail sectors. The scale on which these partners work must fit in with the small-scale volumes of production in the Northern Friesian Woods, or alternatively large-scale processors such as FrieslandCampina would need to be prepared to include 'specialties' in their production lines.

Development of a financial transfer mechanism

The business cases of the Northern Friesian Woods involve the conversion of values. The value of the landscape, for example, is converted into recreational activities, and the value of wooded bank management into energy. The income from these value chains is realised at the end of those chains: by the catering industry/recreation industry entrepreneurs and the energy producers, and not by the farmers.

For these value chains to work effectively, transfer mechanisms therefore need to be developed for sharing the proceeds over the entire value chain. In the case of 'Energy

from wood' this can be handled directly via agreements to do with purchasing, sales and marketing. When it comes to landscape management and recreation things get more complicated. Here, a part of the income from recreation must find its way to the farmers for the purposes of landscape management and the construction and management of paths, etc. Possible instruments in this regard include a tourist tax, regional savings accounts or contracts.

Scaling up dairy farms in combination with other activities

The future developments within the European dairy industry will mean that a dairy farm with 40-50 cows will no longer be commercially viable. The small-scale nature of the landscape in the Northern Friesian Woods is at variance with the growing mechanisation that will result from the increase in farm size.

The results of the studies into increases in scale in a small-scale landscape will need to be implemented in the coming years in the Northern Friesian Woods.

Furnishing proof

The claims of 'healthy milk' will need to be investigated and substantiated. In order to obtain a definitive government commitment in favour of the system of closed-loop farming and landscape management, indicators and monitoring systems will need to be set up.

IN THE NORTHERN FRIESIAN WOODS THERE IS A STRONG SENSE OF COHESION

Members of the Board of the Association of Northern Friesian Woods. From left to right: Foppe Nijboer, Gjalt Benedictus, Dick Zeinstra, Jan Brandsma, Douwe Hoogland and Folkert Algra. (Photo: Mugmedia, Wageningen)

Thanks to...

Mariëtte van Amstel (VU University-Athena Institute)   Ruud Beekhuis (Evelop)   Alfons Beldman (Wageningen UR-LEI)   Arie de Bode (Heros Sluiskil B.V.)   Jos Bode (Evelop)   Gerard Brandsen (poultry-farmer)   Jan Broeze (Wageningen UR-Food & Biobased Research)   Anneke Brouwers (Municipality of Emmen)   Tjard de Cock Buning (VU University-Athena Institute)   Bianca Domhof (LTO North Projects)   Harma Drenth (Boerentaal)   Marco Duineveld (MijnBoer B.V.)   Gerard van Drooge (Projects LTO North)   Harm Jan van Dijk (Landmarkt B.V.)   Dick Engelhardt (Zeeland Seaports)   Paul Galama (Wageningen UR-Livestock Research)   Alwin Gerritsen (Wageningen UR-Alterra)   Peter Groot Koerkamp (Wageningen UR-Livestock Research)   Gao Guihua (Shanghai Industrial Investment Corporation)   Jan Hassink (Wageningen UR-PRI)   Brigitte Hendrikse-Troost (Delta N.V.)   Jaap Hoek Spaans (Landzijde)   Anne Hoes (VU University-Athena Institute)   Henri Holster (Wageningen UR-Livestock Research)   Douwe Hoogland (NFW Chairman, dairy farmer)   Martin Houben (Houbensteyn Group)   Maikki Huurdeman (Van de Bunt Adviseurs)   Arthur Kalkhoven (ZLTO)   Laurens Klerkx (Wageningen UR-Social Sciences)   Sylvia Koenders (Projects LTO North)   Martin van Koppen (HeadVenture)   Marcel Kuijpers (Kuijpers Kip)   Lan Ge (Wageningen UR-LEI)   Wendy Laverman (Wageningen UR-LEI)   Renée Liesveld (VU University-Athena Institute)   Sjef van der Lubbe (Province of Friesland)   Madeleine van Mansfeld (Wageningen UR-Alterra)

André Meekes (Waste Management Middle East)   Trudy van Megen (Knowhouse B.V., now ZLTO)   Wilfried Nielen (Ecoservice Europe B.V.) Ate Oostra (Oostra Consult)   Toine Poppelaars (Province of Zeeland) Cathrien Posthumus (Province of Drenthe)   Ruud Pothoven (Knowhouse B.V.)   Bram Prins (Wageningen UR-LEI)   Barbara Regeer (VU University-Athena Institute)   Jan Willem van der Schans (Wageningen UR-LEI and Social Sciences)   Ad-Willem Schilperoort (MeraPeak)   Yvon Schuler (Orgyd)   Peter Smeets (Wageningen UR-Alterra)   Han Soethoudt (Wageningen UR-Food & Biobased Research)   Sjaak Swart (Groningen University-Science & Society Group)   Johan Vermeulen (Rondeel B.V.) Izak Vermeij (Wageningen UR-Livestock Research)   Peter Vingerling (T&S Consult)   Huub Vousten (Christiaens Group)   Gert-Jan Vullings (pig farmer)   Rinus van de Waart (Knowhouse B.V.)   Adhemar van Waes (Municipality of Terneuzen)   Mark van Waes (Van de Bunt Adviseurs) Berend Jan Wilms (dairy farmer)   Bouke Durk Wilms (dairy farmer) Henny van der Windt (Groningen University-Science & Society Group) Mariët de Winter (Wageningen UR-LEI)   Ruud Zanders (Rondeel B.V.)

And many others who were involved in the TransForum projects!